Thinking

ALSO BY JOHN BROCKMAN

AS AUTHOR

AS EDITOR

AS COEDITOR

Thinking

The New Science of Decision-Making,
Problem-Solving, and Prediction

Edited by John Brockman

HARPER PERENNIAL

NEW YORK • LONDON • TORONTO • SYDNEY • NEW DELHI • AUCKLAND

HARPER PERENNIAL

HarperCollins books may be purchased for educational, business, or sales promotional use. For information please e-mail the Special Markets Department at SPsales@harpercollins.com.

FIRST EDITION

Designed by Michael Correy

Library of Congress Cataloging-in-Publication Data has been applied for.

ISBN 978-0-06-225854-0

18 19 20 ov/LSC 10 9 8 7 6

Contents

PUBLISHER'S NOTE

The essays contained in this book are unedited transcriptions of scientific talks and conversations between conference participants. Any apparent errors in usage should be considered natural products of speech.

1
The Normal Well-Tempered Mind

Daniel C. Dennett

Philosopher; Austin B. Fletcher Professor of Philosophy and Codirector of the Center for Cognitive Studies, Tufts University; author, *Darwin's Dangerous Idea*, *Breaking the Spell*, and *Intuition Pumps*.

I'm trying to undo a mistake I made some years ago, and rethink the idea that the way to understand the mind is to take it apart into simpler minds and then take those apart into still simpler minds until you get down to minds that can be replaced by a machine. This is called homuncular functionalism, because you break the whole person down into two or three or four or seven subpersons who are basically agents. They're homunculi, and this looks like a regress, but it's only a finite regress, because you take each of those in turn and you break it down into a group of stupider, more specialized homunculi, and keep going until you arrive at parts that you can replace with a machine, and that's a great way of thinking about cognitive science. It's what good old-fashioned AI tried to do and is still trying to do.

The idea is basically right, but when I first conceived of it, I made a big mistake. I was at that point enamored of the McCulloch-Pitts logical neuron. McCulloch and Pitts had put together the idea of a very simple artificial neuron, a computational neuron, which had multiple inputs and a single branching output and a threshold for firing, and the inputs were either inhibitory or excitatory. They

1

proved that in principle a neural net made of these logical neurons could compute anything you wanted to compute. So this was very exciting. It meant that basically you could treat the brain as a computer and treat the neuron as a sort of basic switching element in the computer, and that was certainly an inspiring oversimplification. Everybody knew it was an oversimplification, but people didn't realize how much, and more recently it's become clear to me that it's a dramatic oversimplification, because each neuron, far from being a simple logical switch, is a little agent with an agenda, and they are much more autonomous and much more interesting than any switch.

The question is, what happens to your ideas about computational architecture when you think of individual neurons not as dutiful slaves or as simple machines but as agents that have to be kept in line and properly rewarded and that can form coalitions and cabals and organizations and alliances? This vision of the brain as a sort of social arena of politically warring forces seems like sort of an amusing fantasy at first, but is now becoming something that I take more and more seriously, and it's fed by a lot of different currents.

Evolutionary biologist David Haig has some lovely papers on intrapersonal conflicts where he's talking about how even at the level of the genetics—even at the level of the conflict between the genes you get from your mother and the genes you get from your father, the so-called madumnal and padumnal genes—those are in opponent relations, and if they get out of whack, serious imbalances can happen that show up as particular psychological anomalies.

We're beginning to come to grips with the idea that your brain is not this well-organized hierarchical control system where everything is in order, a very dramatic vision of bureaucracy. In fact,

Daniel C. Dennett

it's much more like anarchy with some elements of democracy. Sometimes you can achieve stability and mutual aid and a sort of calm united front, and then everything is hunky-dory, but then it's always possible for things to get out of whack and for one alliance or another to gain control, and then you get obsessions and delusions and so forth.

You begin to think about the normal well-tempered mind, in effect, the well-organized mind, as an achievement, not as the base state, something that is only achieved when all is going well. But still, in the general realm of humanity, most of us are pretty well put together most of the time. This gives a very different vision of what the architecture is like, and I'm just trying to get my head around how to think about that.

What we're seeing right now in cognitive science is something that I've been anticipating for years, and now it's happening, and it's happening so fast I can't keep up with it. We're now drowning in data, and we're also happily drowning in bright young people who have grown up with this stuff and for whom it's just second nature to think in these quite abstract computational terms, and it simply wasn't possible even for experts to get their heads around all these different topics 30 years ago. Now a suitably motivated kid can arrive at college already primed to go on these issues. It's very exciting, and they're just going to run away from us, and it's going to be fun to watch.

The vision of the brain as a computer, which I still champion, is changing so fast. The brain's a computer, but it's so different from any computer that you're used to. It's not like your desktop or your laptop at all, and it's not like your iPhone, except in some ways. It's a much more interesting phenomenon. What Turing gave us for the first time (and without Turing you just couldn't do any of this) is a way of thinking in a disciplined way about phenomena that have, as I like to say, trillions of moving parts. Until the late 20th century, nobody knew how to take seriously a machine with a trillion moving parts. It's just mind-boggling.

You couldn't do it, but computer science gives us the ideas, the concepts of levels—virtual machines implemented in virtual machines implemented in virtual machines and so forth. We have these nice ideas of recursive reorganization of which your iPhone is just one example, and a very structured and very rigid one, at that.

We're getting away from the rigidity of that model, which was worth trying for all it was worth. You go for the low-hanging fruit first. First, you try to make minds as simple as possible. You make them as much like digital computers, as much like von Neumann machines, as possible. It doesn't work. Now, we know why it doesn't work pretty well. So you're going to have a parallel architecture because, after all, the brain is obviously massively parallel.

It's going to be a connectionist network. Although we know many of the talents of connectionist networks, how do you knit them together into one big fabric that can do all the things minds do? Who's in charge? What kind of control system? Control is the real key, and you begin to realize that control in brains is very different from control in computers. Control in your commercial computer is very much a carefully designed top-down thing.

You really don't have to worry about one part of your laptop going rogue and trying out something on its own that the rest of the system doesn't want to do. No, they're all slaves. If they're agents, they're slaves. They are prisoners. They have very clear job descriptions. They get fed every day. They don't have to worry about where the energy's coming from, and they're not ambitious. They just do what they're asked to do, and they do it brilliantly, with only the slightest tint of comprehension. You get all

Daniel C. Dennett

the power of computers out of these mindless little robotic slave prisoners, but that's not the way your brain is organized.

Each neuron is imprisoned in your brain. I now think of these as cells within cells, as cells within prison cells. Realize that every neuron in your brain, every human cell in your body (leaving aside all the symbionts), is a direct descendant of eukaryotic cells that lived and fended for themselves for about a billion years as free-swimming, free-living little agents. They fended for themselves, and they survived.

They had to develop an awful lot of know-how, a lot of talent, a lot of self-protective talent to do that. When they joined forces into multicellular creatures, they gave up a lot of that. They became, in effect, domesticated. They became part of larger, more monolithic organizations. My hunch is that that's true in general. We don't have to worry about our muscle cells rebelling against us, or anything like that. When they do, we call it cancer, but in the brain I think that (and this is my wild idea) maybe only in one species, us, and maybe only in the obviously more volatile parts of the brain, the cortical areas, some little switch has been thrown in the genetics that, in effect, makes our neurons a little bit feral, a little bit like what happens when you let sheep or pigs go feral, and they recover their wild talents very fast.

Maybe a lot of the neurons in our brains are not just capable but, if you like, motivated to be more adventurous, more exploratory or risky in the way they comport themselves, in the way they live their lives. They're struggling among themselves with each other for influence, just for staying alive, and there's competition going on between individual neurons. As soon as that happens, you have room for cooperation to create alliances, and I suspect that a more free-wheeling, anarchic organization is the secret of our greater capacities of creativity, imagination, thinking outside the box and all that, and the price we pay for it is our susceptibility to obsessions, mental illnesses, delusions, and smaller problems.

We got risky brains that are much riskier than the brains of other mammals, even more risky than the brains of chimpanzees, and this could be partly a matter of a few simple mutations in control genes that release some of the innate competitive talent that is still there in the genomes of the individual neurons. But I don't think that genetics is the level to explain this. You need culture to explain it.

This, I speculate, is a response to our invention of culture; culture creates a whole new biosphere, in effect, a whole new cultural sphere of activity where there's opportunities that don't exist for any other brain tissues in any other creatures, and that this exploration of this space of cultural possibility is what we need to do to explain how the mind works.

Everything I just said is very speculative. I'd be thrilled if 20 percent of it was right. It's an idea, a way of thinking about brains and minds and culture that is, to me, full of promise, but it may not pan out. I don't worry about that, actually. I'm content to explore this, and if it turns out that I'm just wrong, I'll say, "Oh, okay. I was wrong. It was fun thinking about it." But I think I might be right.

I'm not myself equipped to work on a lot of the science; other people could work on it, and they already are, in a way. The idea of selfish neurons has already been articulated by Sebastian Seung of MIT in a brilliant keynote lecture he gave at the Society for Neuroscience in San Diego a few years ago. I thought, Oh, yeah, selfish neurons, selfish synapses. Cool. Let's push that and see where it leads. But there are many ways of exploring this. One of the still unexplained, so far as I can tell, and amazing features of the brain is its tremendous plasticity.

Mike Merzenich sutured a monkey's fingers together so that it didn't need as much cortex to represent two separate individual digits, and pretty soon the cortical regions that were representing those two digits shrank, making that part of the cortex available to use for other things. When the sutures were removed, the cortical regions soon resumed pretty much their earlier dimensions. If you blindfold yourself for eight weeks, as Alvaro Pascual-Leone does in his experiments, you find that your visual cortex starts getting adapted for Braille, for haptic perception, for touch.

The way the brain spontaneously reorganizes itself in response to trauma of this sort, or just novel experience, is itself one of the most amazing features of the brain, and if you don't have an architecture that can explain how that could happen and why that is, your model has a major defect. I think you really have to think in terms of individual neurons as micro-agents, and ask what's in it for them.

Why should these neurons be so eager to pitch in and do this other work just because they don't have a job? Well, they're out of work. They're unemployed, and if you're unemployed, you're not getting your neuromodulators. If you're not getting your neuromodulators, your neuromodulator receptors are going to start disappearing, and pretty soon you're going to be really out of work, and then you're going to die.

In this regard, I think of John Holland's work on the emergence of order. His example is New York City. You can always find a place where you can get gefilte fish, or sushi, or saddles, or just about anything under the sun you want, and you don't have to worry about a state bureaucracy that is making sure that supplies get through. No. The market takes care of it. The individual web of entrepreneurship and selfish agency provides a host of goods and services, and is an extremely sensitive instrument that responds to needs very quickly.

Until the lights go out. Well, we're all at the mercy of the power man. I am quite concerned that we're becoming hyper-fragile as a civilization, and we're becoming so dependent on technologies that are not as reliable as they should be, that have so many conditions that have to be met for them to work, that we may specialize ourselves into some very serious jams. But in the meantime, thinking about the self-organizational powers of the brain as being very much like the self-organizational powers of a city is not a bad idea. It just reeks of overenthusiastic metaphor, though, and it's worth reminding ourselves that this idea has been around since Plato.

Plato analogizes the mind of a human being to the state. You've got the rulers and the guardians and the workers. This idea that a person is made of lots of little people is comically simpleminded in some ways, but that doesn't mean it isn't, in a sense, true. We shouldn't shrink from it just because it reminds us of simpleminded versions that have been long discredited. Maybe some not-so-simpleminded version is the truth.

There are a lot of cultural fleas

My next major project will be trying to take another hard look at cultural evolution and look at the different views of it and see if I can achieve a sort of bird's-eye view and establish what role, if any, there is for memes or something like memes, and what the other forces are that are operating. We are going to have to have a proper scientific perspective on cultural change. The old-fashioned, historical narratives are wonderful, and they're full of gripping detail, and they're even sometimes right, but they only cover a small proportion of the phenomena. They only cover the tip of the iceberg.

Basically, the model that we have and have used for several

　　　　　　　　　　　　　　Daniel C. Dennett

thousand years is the model that culture consists of treasures, cultural treasures. Just like money, or like tools and houses, you bequeath them to your children, and you amass them, and you protect them, and because they're valuable, you maintain them and prepare them, and then you hand them on to the next generation, and some societies are rich, and some societies are poor, but it's all goods. I think that vision is true of only the tip of the iceberg.

Most of the regularities in culture are not treasures. It's not all opera and science and fortifications and buildings and ships. It includes all kinds of bad habits and ugly patterns and stupid things that don't really matter but that somehow have got a grip on a society and that are part of the ecology of the human species, in the same way that mud, dirt and grime, and fleas are part of the world that we live in. They're not our treasures. We may give our fleas to our children, but we're not trying to. It's not a blessing. It's a curse, and I think there are a lot of cultural fleas. There are lots of things that we pass on without even noticing that we're doing it, and, of course, language is a prime case of this—very little deliberate, intentional language instruction goes on or has to go on.

Kids that are raised with parents pointing out individual objects and saying, "See, it's a ball. It's red. Look, Johnny, it's a red ball, and this is a cow, and look at the horsy" learn to speak, but so do kids who don't have that patient instruction. You don't have to do that. Your kids are going to learn *ball* and *red* and *horsy* and *cow* just fine without that, even if they're quite severely neglected. That's not a nice observation to make, but it's true. It's almost impossible not to learn language if you don't have some sort of serious pathology in your brain.

Compare that with chimpanzees. There are hundreds of chimpanzees who have spent their whole lives in human captivity.

They've been institutionalized. They've been like prisoners, and in the course of the day they hear probably about as many words as a child does. They never show any interest. They apparently never get curious about what those sounds are for. They can hear all the speech, but it's like the rustling of the leaves. It just doesn't register on them as being worth attention.

But kids are tuned for that, and it might be a very subtle tuning. I can imagine a few small genetic switches that, if they were just in a slightly different position, would make chimpanzees just as pantingly eager to listen to language as human babies are—but they're not, and what a difference it makes in their world! They never get to share discoveries the way we do, or share our learning. That, I think, is the single feature about human beings that distinguishes us most clearly from all others: we don't have to reinvent the wheel. Our kids get the benefit of not just what grandpa and grandma and great-grandpa and great-grandma knew. They get the benefit of basically what everybody in the world knew, in the years when they go to school. They don't have to invent calculus or long division or maps or the wheel or fire. They get all that for free. It just comes as part of the environment. They get incredible treasures, cognitive treasures, just by growing up.

I've got a list as long as my arm of stuff that I've been trying to get time to read. I'm going to Paris in December and talking at the Dan Sperber conference, and I'm going to be addressing Dan's concerns about cultural evolution. I think he's got some great ideas and some ideas I think he's wrong about. So that's a very fruitful disagreement for me.

Daniel C. Dennett

A lot of naïve thinking by scientists about free will

"Moving Naturalism Forward" was a nice workshop that Sean Carroll put together out in Stockbridge a couple of weeks ago, and it was really interesting. I learned a lot. I learned more about how hard it is to do some of these things, and that's always useful knowledge, especially for a philosopher.

If we take seriously, as I think we should, the role that Socrates proposed for us as midwives of thinking, then we want to know what the blockades are, what the imagination blockades are, what people have a hard time thinking about—and among the things that struck me about the Stockbridge conference were the signs of people really having to struggle to take seriously some ideas that I think they should take seriously.

I was struggling, too, because there were scientific ideas that I found hard to get my head around. It's interesting that you can have a group of people who are trying to communicate. They're not showing off. They're interested in finding points of common agreement, and they're still having trouble, and that's something worth seeing and knowing what that's about, because then you go into the rest of your forays sadder but wiser. Well, sort of. You at least are alert to how hard it can be to implant a perspective or a way of thinking in somebody else's mind.

I realized I really have my work cut out for me in a way that I had hoped not to discover. There's still a lot of naïve thinking by scientists about free will. I've been talking about it quite a lot, and I do my best to undo some bad thinking by various scientists. I've had some modest success, but there's a lot more that has to be done on that front. I think it's very attractive to scientists to think that here's this several-millennia-old philosophical idea, free will, and they can just hit it out of the ballpark, which I'm sure would be nice if it was true.

It's just not true. I think they're well intentioned. They're trying to clarify, but they're really missing a lot of important points. I want a naturalistic theory of human beings and free will and moral responsibility as much as anybody there, but I think you've got to think through the issues a lot better than they've done, and this, happily, shows that there's some real work for philosophers.

Philosophers have done some real work that the scientists jolly well should know. Here's an area where it was one of the few times in my career when I wanted to say to a bunch of scientists, "Look. You have some reading to do in philosophy before you hold forth on this. There really is some good reading to do on these topics, and you need to educate yourselves."

A combination of arrogance and cravenness

The figures about American resistance to evolution are still depressing, and you finally have to realize that there's something structural. It's not that people are stupid, and I think it's clear that people, everybody, me, you, we all have our authorities, our go-to people whose word we trust. If you want to ask a question about the economic situation in Greece, for instance, you need to check it out with somebody whose opinion on that is worth taking seriously. We don't try to work it out for ourselves. We find some expert that we trust, and right around the horn, whatever the issues are, we have our experts. A lot of people have their pastors as their experts on matters of science. This is their local expert.

I don't blame them. I wish they were more careful about vetting their experts and making sure that they found good experts. They

Daniel C. Dennett

wouldn't choose an investment adviser, I think, as thoughtlessly as they go along with their pastor. I blame the pastors, but where do they get their ideas? Well, they get them from the hierarchies of their churches. Where do they get their ideas? Up at the top, I figure there's some people that really should be ashamed of themselves. They know better.

They're lying, and when I get a chance, I try to ask them that. I say, "Doesn't it bother you that your grandchildren are going to want to know why you thought you had to lie to everybody about evolution?" I mean, really. They're lies. They've got to know that these are lies. They're not that stupid, and I just would love them to worry about what their grandchildren and great-grandchildren would say about how their ancestors were so craven and so arrogant. It's a combination of arrogance and cravenness.

We now have to start working on that structure of experts and thinking, why does that persist? How can it be that so many influential, powerful, wealthy, in-the-public people can be so confidently wrong about evolutionary biology? How did that happen? Why does it happen? Why does it persist? It really is a bit of a puzzle, if you think about how embarrassed they'd be not to know that the world is round. I think it would be deeply embarrassing to be that benighted, and they'd realize it. They'd be embarrassed not to know that HIV is the vector of AIDS. They'd be embarrassed to not understand the way the tides are produced by the gravitational forces of the moon and the sun. They may not know the details, but they know that the details are out there. They could learn them in 20 minutes if they wanted to. How did they get themselves in the position where they could so blithely trust people who they'd never buy stocks and bonds from? They'd never trust a child's operation to a doctor who was as ignorant and as ideological as these people. It is really strange. I haven't gotten to the bottom of that.

This pernicious sort of lazy relativism

A few years ago, Linda LaScola, who's a very talented investigator, questioner, and interviewer, and I started a project where we found closeted nonbelieving pastors who still had churches and would speak in confidence to her. She's a very good interviewer, and she got and earned their trust, and then they really let their hair down and explained how they got in the position they're in and what it's like. What is it like to be a pastor who has to get up and say the creed every Sunday when you don't believe that anymore? And they're really caught in a nasty trap.

When we published the first study, there was a lot of reaction, and one of the amazing things was the dogs that didn't bark. Nobody said we were making it up or it wasn't a problem. Every religious leader knows. It's their dirty little secret. They knew jolly well that what we were looking at was the tip of an iceberg, that there are a lot of pastors out there who simply don't believe what their parishioners think they believe, and some of them are really suffering, and some of them aren't, and that's interesting, too.

In phase two we've spread out and looked at a few more, and we've also started looking at seminary professors, the people that teach the pastors what they learn and often are instrumental in starting them down the path of this sort of systematic hypocrisy where they learn in seminary that there's what you can talk about in the seminary, and there's what you can say from the pulpit, and those are two different things. I think this phenomenon of systematic hypocrisy is very serious. It is the structural problem in religion today, and churches deal with it in various ways, none of them very good.

The reason they can't deal with them well is that they have a principle, which is a little bit like the Hippocratic oath of medi-

Daniel C. Dennett

cine: First, do no harm. Well, they learn this, and they learn that from the pulpit the one thing they mustn't do is shake anybody's faith. If they've got a parish full of literalists, young earth creationists, literal Bible believers who believe that all the miracles in the Bible really happened and that the resurrection is the literal truth and all that, they must not disillusion those people. But then they also realize that a lot of other parishioners are not so sure; they think it's all sort of metaphor—symbolic, yes, but they don't take it as literally true.

How do they thread the needle so that they don't offend the sophisticates in their congregation by insisting on the literal truth of the book of Genesis, let's say, while still not scaring, betraying, pulling the rug out from under the more naïve and literal-minded of their parishioners? There's no good solution to that problem as far as we can see, since they have this unspoken rule that they should not upset, undo, subvert the faith of anybody in the church.

This means there's a sort of enforced hypocrisy in which the pastors speak from the pulpit quite literally, and if you weren't listening very carefully, you'd think: oh my gosh, this person really believes all this stuff. But they're putting in just enough hints for the sophisticates in the congregation so that the sophisticates are supposed to understand: Oh, no. This is all just symbolic. This is all just metaphorical. And that's the way they want it, but of course they could never admit it. You couldn't put a little neon sign up over the pulpit that says, "Just metaphor, folks, just metaphor." It would destroy the whole thing.

You can't admit that it's just metaphor even when you insist when anybody asks that it's just metaphor, and so this professional doubletalk persists, and if you study it for a while the way Linda and I have been doing, you come to realize that's what it is, and that means they've lost track of what it means to tell the truth.

Oh, there are so many different kinds of truth. Here's where post-modernism comes back to haunt us. What a pernicious bit of intellectual vandalism that movement was! It gives license to this pernicious sort of lazy relativism.

One of the most chilling passages in that great book by William James, *The Varieties of Religious Experience*, is where he talks about soldiers in the military: "Far better is it for an army to be too savage, too cruel, too barbarous, than to possess too much sentimentality and human reasonableness." This is a very sobering, to me, a very sobering reflection. Let's talk about when we went into Iraq. There was Rumsfeld saying, "Oh, we don't need a big force. We don't need a big force. We can do this on the cheap," and there were other people—retrospectively, we can say they were wiser—who said, "Look, if you're going to do this at all, you want to go in there with such overpowering, such overwhelming numbers and force that you can really intimidate the population, and you can really maintain the peace and just get the population to sort of roll over, and that way actually less people get killed, less people get hurt. You want to come in with an overwhelming show of force."

We didn't do that, and look at the result. Terrible. Maybe we couldn't do it. Maybe Rumsfeld knew that the American people would never stand for it. Well, then, they shouldn't go in, because look what happened. But the principle is actually one that's pretty well understood. If you don't want to have a riot, have four times more police there than you think you need. That's the way not to have a riot and nobody gets hurt, because people are not foolish enough to face those kinds of odds. But they don't think about that with regard to religion, and it's very sobering.

I put it this way. Suppose that we face some horrific, terrible enemy, another Hitler or something really, really bad, and here's two different armies that we could use to defend ourselves. I'll

Daniel C. Dennett

call them the Gold Army and the Silver Army: same numbers, same training, same weaponry. They're all armored and armed as well as we can do. The difference is that the Gold Army has been convinced that God is on their side and this is the cause of righteousness, and it's as simple as that. The Silver Army is entirely composed of economists. They're all making side insurance bets and calculating the odds of everything.

Which army do you want on the front lines? It's very hard to say you want the economists, but think of what that means. What you're saying is that we'll just have to hoodwink all these young people into some false beliefs for their own protection and for ours. It's extremely hypocritical. It is a message that I recoil from, the idea that we should indoctrinate our soldiers. In the same way that we inoculate them against diseases, we should inoculate them against the economists'—or philosophers'—sort of thinking, since it might lead them to think: Am I so sure this cause is just? Am I really prepared to risk my life to protect? Do I have enough faith in my commanders that they're doing the right thing? What if I'm clever enough and thoughtful enough to figure out a better battle plan, and I realize that this is futile? Am I still going to throw myself into the trenches? It's a dilemma that I don't know what to do about, although I think we should confront it, at least.

2

How to Win at Forecasting

Philip Tetlock

Leonore Annenberg University Professor of Psychology,
University of Pennsylvania; author, *Expert Political Judgment*.

INTRODUCTION by Daniel Kahneman
Recipient of the 2002 Nobel Prize in Economics; Eugene Higgins
Professor of Psychology Emeritus, Princeton University; author,
Thinking, Fast and Slow.

Philip Tetlock's 2005 book Expert Political Judgment: How Good
Is It? How Can We Know? *demonstrated that accurate long-term
political forecasting is, to a good approximation, impossible. The work
was a landmark in social science, and its importance was quickly recognized and rewarded in two academic disciplines—political science and
psychology. Perhaps more significantly, the work was recognized in the
intelligence community, which accepted the challenge of investing significant resources in a search for improved accuracy. The work is ongoing,
important discoveries are being made, and Tetlock gives us a chance to
peek at what is happening.*

*Tetlock's current message is far more positive than was his earlier dis-
mantling of long-term political forecasting. He focuses on the near term,
where accurate prediction is possible to some degree, and he takes on the
task of making political predictions as accurate as they can be. He has successes to report. As he points out in his comments, these successes will be
destabilizing to many institutions, in ways both multiple and profound.*

With some confidence, we can predict that another landmark of applied social science will soon be reached.

There's a question that I've been asking myself for nearly three decades now and trying to get a research handle on, and that is: why is the quality of public debate so low, and why is it that the quality often seems to deteriorate the more important the stakes get?

About 30 years ago I started my work on expert political judgment. It was the height of the Cold War. There was a ferocious debate about how to deal with the Soviet Union. There was a liberal view; there was a conservative view. Each position led to certain predictions about how the Soviets would be likely to react to various policy initiatives.

One thing that became very clear, especially after Gorbachev came to power and confounded the predictions of both liberals and conservatives, was that even though nobody predicted the direction that Gorbachev was taking the Soviet Union, virtually everybody after the fact had a compelling explanation for it. We seemed to be working in what one psychologist called an "outcome-irrelevant learning situation." People drew whatever lessons they wanted from history.

There is quite a bit of skepticism about political punditry, but there's also a huge appetite for it. I was struck 30 years ago and I'm struck now by how little interest there is in holding political pundits who wield great influence accountable for predictions they make on important matters of public policy.

The presidential election of 2012, of course, brought about the Nate Silver controversy, and a lot of people, mostly Democrats, took great satisfaction out of Silver being more accurate than leading Republican pundits. It's undeniably true that he was more accurate. He was using more rigorous techniques in analyzing and aggregating data than his competitors and debunkers were.

But it's not something uniquely closed-minded about conservatives that caused them to dislike Silver. When you go back to presidential elections that Republicans won, it's easy to find commentaries in which liberals disputed the polls and complained that the polls were biased. That was true even in a blowout political election like 1972, the McGovern-Nixon election. There were some liberals who had convinced themselves that the polls were profoundly inaccurate. It's easy for partisans to believe what they want to believe, and political pundits are often more in the business of bolstering the prejudices of their audience than they are in trying to generate accurate predictions of the future.

Thirty years ago we started running some very simple forecasting tournaments, and they gradually expanded. We were interested in answering a very simple question, and that is what, if anything, distinguishes political analysts who are more accurate from those who are less accurate on various categories of issues. We looked hard for correlates of accuracy. We were also interested in the prior question of whether political analysts can do appreciably better than chance.

We found two things. One, it's very hard for political analysts to do appreciably better than chance when you move beyond about one year. Second, political analysts think they know a lot more about the future than they actually do. When they say they're 80 or 90 percent confident, they're often right only 60 or 70 percent of the time.

There was systematic overconfidence. Moreover, political analysts were disinclined to change their minds when they got it wrong. When they made strong predictions that something was going to

　　　　　　　　　　　　　　　　　　　　　　　Philip Tetlock

happen and it didn't, they were inclined to argue something along the lines of, "Well, I predicted that the Soviet Union would continue, and it would have if the coup plotters against Gorbachev had been more organized," or "I predicted that Canada would disintegrate or Nigeria would disintegrate and it's still there, but it's just a matter of time before it disappears," or "I predicted that the Dow would be down 36,000 by the year 2000 and it's going to get there eventually, but it will just take a bit longer."

So we found three basic things: many pundits were hard-pressed to do better than chance, were overconfident, and were reluctant to change their minds in response to new evidence. That combination doesn't exactly make for a flattering portrait of the punditocracy.

We did a book in 2005, and it's been quite widely discussed. Perhaps the most important consequence of publishing the book is that it encouraged some people within the U.S. intelligence community to start thinking seriously about the challenge of creating accuracy metrics and for monitoring how accurate analysts are—which has led to the major project that we're involved in now, sponsored by the Intelligence Advanced Research Projects Activities (IARPA). It extends from 2011 to 2015, and involves thousands of forecasters making predictions on hundreds of questions over time and tracking in accuracy.

Exercises like this are really important for a democracy. The Nate Silver episode illustrates in a small way what I hope will happen over and over again over the next several decades, which is that there are ways of benchmarking the accuracy of pundits. If pundits feel that their accuracy is benchmarked, they will be more careful and thoughtful about what they say, and it will elevate the quality of public debate.

One of the reactions to my work on expert political judgment was that it was politically naïve; I was assuming that political analysts were in the business of making accurate predictions, whereas they're really in a different line of business altogether. They're in the business of flattering the prejudices of their base audience and entertaining their base audience, and accuracy is a side constraint. They don't want to be caught making an overt mistake, so they generally are pretty skillful in avoiding being caught by using vague verbiage to disguise their predictions. They don't say there's a .7 likelihood of a terrorist attack within this span of time. They don't say there's a 1.0 likelihood of recession by the third quarter of 2013. They don't make predictions like that. What they say is that if we go ahead with the administration's proposed tax increase, there could be a devastating recession in the next six months. "There could be."

The word "could" is notoriously ambiguous. When you ask research subjects what "could" means, it depends enormously on the context: "we could be struck by an asteroid in the next 25 seconds," which people might interpret as something like a .0000001 probability, or "this really could happen," which people might interpret as a .6 or .7 probability. It depends a lot on the context. Pundits have been able to insulate themselves from accountability for accuracy by relying on vague verbiage. They can often be wrong, but never in error.

There is an interesting case study to be done on the reactions of the punditocracy to Silver. Those who are most upfront in debunking him, holding him in contempt, ridiculing him, and offering contradictory predictions were put in a genuinely awkward situation because they were so flatly disconfirmed. They had violated one of the core rules of their own craft, which is to insulate themselves in vague verbiage—to say, "Well, it's possible that

Philip Tetlock

Obama would win." They should have cushioned themselves in various ways with rhetoric.

How do people react when they're actually confronted with error? You get a huge range of reactions. Some people just don't have any problem saying, "I was wrong. I need to rethink this or that assumption." Generally, people don't like to rethink really basic assumptions. They prefer to say, "Well, I was wrong about how good Romney's get-out-the-vote effort was." They prefer to tinker with the margins of their belief system (e.g., "I fundamentally misread U.S. domestic politics, my core area of expertise").

A surprising fraction of people are reluctant to acknowledge there was anything wrong with what they were saying. One argument you sometimes hear—and we heard this in the abovementioned episode, but I also heard versions of it after the Cold War—is, "I was wrong, but I made the right mistake." Dick Morris, the Republican pollster and analyst, conceded that he was wrong, but it was the right mistake to make because he was acting, essentially, as a cheerleader for a particular side and it would have been far worse to have underestimated Romney than to have overestimated him.

If you have a theory how world politics works that can lead you to value avoiding one error more than the complementary error, you might say, "Well, it was really important to bail out this country because if we hadn't, it would have led to financial contagion. There was a risk of losing our money in the bailout, but the risk was offset because I thought the risk of contagion was substantial." If you have a contagion theory of finance, that theory will justify putting bailout money at risk. If you have a theory that the enemy is only going to grow bolder if you don't act really strongly against it, then you're going to say, "Well, the worst mistake would have been to appease them, so we hit them really hard. And even though that led to an expansion of the conflict, it would have been far worse if we'd gone down the other path." It's very, very hard to pin them down, and that's why these types of level-playing-field forecasting tournaments can play a vital role in improving the quality of public debate.

There are various interesting scientific objections that have been raised to these level-playing-field forecasting exercises. One line of objection would be grounded more in Nassim Taleb's school of thought, the black swan view of history: where we are in history today is the product of forces that not only no one foresaw, but no one could have foreseen. The epoch-transforming events like World War One, nuclear bombs and nuclear missiles to deliver them, and the invention of the Internet—these are geopolitical and technological transformational events in history no one foresaw, no one could have foreseen. In this view, history is best understood in tems of a punctuated equilibrium model. There are periods of calm and predictability punctuated by violent exogenous shocks that transform things—sometimes for the better, sometimes for the worse—and these discontinuities are radically unpredictable.

What are we doing? Well, in this view, we may be lulling people into a kind of false complacency by giving them the idea that you can improve your foresight to an ascertainable degree within ascertainable time parameters and types of tasks. That's going to induce a false complacency and will cause us to be blindsided all the more violently by the next black swan, because we think we have a good probabilistic handle on an erratically unpredictable world—which is an interesting objection, and something we have to be on the lookout for.

There is, of course, no evidence to support that claim. I would argue that making people more appropriately humble about their ability to predict a short-term future is probably, on balance, going to make them more appropriately humble about their abil-

Philip Tetlock

ity to predict the long-term future, but that certainly is a line of argument that's been raised about the tournament.

Another interesting variant of that argument is that it's possible to learn in certain types of tasks, but not in other types of tasks. It's possible to learn, for example, how to be a better poker player. Nate Silver could learn to be a really good poker player. Hedge fund managers tend to be really good poker players, probably because it's good preparation for their job. Well, what does it mean to be a good poker player? You learn to be a good poker player because you get repeated clear feedback and you have a well-defined sampling universe from which the cards are being drawn. You can actually learn to make reasonable probability estimates about the likelihood of various types of hands materializing in poker.

Is world politics like a poker game? This is what, in a sense, we are exploring in the IARPA forecasting tournament. You can make a good case that history is different and poses unique challenges. This is an empirical question of whether people can learn to become better at these types of tasks. We now have a significant amount of evidence on this, and the evidence is that people can learn to become better. It's a slow process. It requires a lot of hard work, but some of our forecasters have really risen to the challenge in a remarkable way and are generating forecasts that are far more accurate than I would have ever supposed possible from past research in this area.

Silver's situation is more like poker than geopolitics. He has access to polls that are being drawn from representative samples. The polls have well-defined statistical properties. There's a well-defined sampling universe, so he is closer to the poker domain when he is predicting electoral outcomes in advanced democracies with well-established polling procedures, well-established sampling methodologies, and relatively uncorrupted polling processes. That's more like poker and less like trying to predict the outcome of a civil war in sub-Saharan Africa or trying to predict that H5N1 is going to spread in a certain way, or many of the types of events that loom large in geopolitical or technological forecasting.

There has long been disagreement among social scientists about how scientific social science can be, and the skeptics have argued that social phenomena are more cloudlike. They don't have Newtonian clocklike regularity. That cloud versus clock distinction has loomed large in those kinds of debates. If world politics were truly clocklike and deterministic then it should, in principle, be possible for an observer who is armed with the correct theory and correct knowledge of the antecedent conditions to predict with extremely high accuracy what's going to happen next.

If world politics is more cloudlike—little wisps of clouds blowing around in the air in quasi-random ways—no matter how theoretically prepared the observer is, the observer is not going to be able to predict very well. Let's say the clocklike view posits that the optimal forecasting frontier is very close to 1.0, an R squared very close to 1.0. By contrast, the cloudlike view would posit that the optimal forecasting frontier is not going to be appreciably greater than chance or you're not going to be able to do much better than a dart-throwing chimpanzee. One of the things that we discovered in the earlier work was that forecasters who suspected that politics was more cloudlike were actually more accurate in predicting longer-term futures than forecasters who believed that it was more clocklike.

Forecasters who were more modest about what could be accomplished predictably were actually generating more accurate predictions than forecasters who were more confident about what could be achieved. We called these theoretically confident forecasters "hedgehogs." We called these more modest, self-critical forecasters "foxes," drawing on Isaiah Berlin's famous essay "The Hedgehog and the Fox."

Let me say something about how dangerous it is to draw strong inferences about accuracy from isolated episodes. Imagine, for example, that Silver had been wrong and that Romney had become president. And let's say his prediction had been a 0.8 probability two weeks prior to the election that made Romney president. You can imagine what would have happened to his credibility. It would have cratered. People would have concluded that, yes, his Republican detractors were right, that he was essentially an Obama hack, and he wasn't a real scientist. That's, of course, nonsense. When you say there's a .8 probability, there's 20 percent chance that something else could happen. And it should reduce your confidence somewhat in him, but you shouldn't abandon him totally. There's a disciplined Bayesian belief adjustment process that's appropriate in response to miscalibrated forecasts.

What we see instead is overreactions. Silver would be a fool if he'd gotten it wrong, or he's a god if he gets it right. He's neither a fool nor a god. He's a thoughtful data analyst who knows how to work carefully through lots of detailed data and aggregate them in sophisticated ways and get a bit of a predictive edge over many, but not all, of his competitors. There are other aggregators out there who are doing as well or maybe even a little bit better, but their methodologies are quite strikingly similar and they're relying on a variant of the wisdom of the crowd, which is aggregation. They're pooling a lot of diverse bits of information and they're trying to give more weight to those bits of information that have a good historical track record of having been accurate. It's a weighted averaging kind of process, essentially, and that's a good strategy.

I don't have a dog in this theoretical fight. There's one school of thought that puts a lot of emphasis on the advantages of "blink," on the advantages of going with your gut. There's another school of thought that puts a lot of emphasis on the value of system-two overrides, self-critical cognition—giving things over a second thought. For me it is really a straightforward empirical question of, what are the conditions under which each style of thinking works better or worse?

In our work on expert political judgment we have generally had a hard time finding support for the usefulness of fast and frugal simple heuristics. It's generally the case that forecasters who are more thoughtful and self-critical do a better job of attaching accurate probability estimates to possible futures. I'm sure there are situations when going with a blink may well be a good idea, and I'm sure there are situations when we don't have time to think. When you think there might be a tiger in the jungle, you might want to move very fast, before you fully process the information. That's all well-known and discussed elsewhere. For us, we're finding more evidence for the value of thoughtful system-two overrides, to use Danny Kahneman's terminology.

Let's go back to this fundamental question of, what are we capable of learning from history, and are we capable of learning anything from history that we weren't already ideologically predisposed to learn? As I mentioned before, history is not a good teacher, and we see what a capricious teacher history is in the reactions to Nate Silver in the 2012 election forecasting—he's either a genius or he's an idiot. And we need to have much more nuanced, well-calibrated reactions to episodes of this sort.

The intelligence community is responsible, of course, for providing the U.S. government with timely advice about events around the world, and they frequently get politically clobbered, virtually whenever they make mistakes. There are two types of mistakes you can make, essentially. You can make a false-positive prediction or you can make a false-negative prediction.

What would a false-positive prediction look like? Well, the most famous recent false-positive prediction is probably the false positive on weapons of mass destruction in Iraq, which led to a trillion-plus-dollar war. What about famous false-negative predictions? Well, a lot of people would call 9/11 a serious false negative. The intelligence community oscillates back and forth in response to these sharp political critiques that are informed by hindsight, and one of the things that we know from elementary behaviorism as well as from work in organizational learning is that rats, people, and organizations do respond to rewards and punishments. If an organization has been recently clobbered for making a false-positive prediction, that organization is going to make major efforts to make sure it doesn't make another false positive. They're going to be so sure that they might make a lot more false negatives in order to avoid that. "We're going to make sure we're not going to make a false positive even if that means we're going to underestimate the Iranian nuclear program." Or "We're going to be really sure we don't make a false negative even if that means we have false alarms of terrorism for the next 25 years."

The question becomes, is it possible to set up a system for learning from history that's not simply programmed to avoid the most recent mistake in a very simple, mechanistic fashion? Is it possible to set up a system for learning from history that actually learns in our sophisticated way that manages to bring down both false positives and false negatives to some degree? That's a big question mark.

Nobody has really systematically addressed that question until IARPA, the Intelligence Advanced Research Projects Activities, sponsored this particular project, which is very, very ambitious in scale. It's an attempt to address the question of whether you can push political forecasting closer to what philosophers might call an optimal forecasting frontier. An optimal forecasting frontier is a frontier along which you just can't get any better. You can't get false positives down anymore without having more false negatives. You can't get false negatives down anymore without having more false positives. That's just the optimal state of prediction, unless you subscribe to an extremely clocklike view of the political, economical, and technological universe. If you subscribe to that, you might believe that the optimal forecasting frontier is 1.0 and that godlike omniscience is possible. You never have to tolerate any false positives or false negatives.

There are very few people on the planet, I suspect, who believe that to be true of our world. But you don't have to go all the way to the cloudlike extreme and say that we are all just radically unpredictable. Most of us are somewhere in between clocklike and cloudlike, but we don't know for sure where we are in that distribution, and IARPA is helping us to figure out where we are.

It's fascinating to me that there is a steady public appetite for books that highlight the feasibility of prediction like Nate Silver, and there's a deep public appetite for books like Nassim Taleb's *The Black Swan*, which highlights the apparent unpredictability of our universe. The truth is somewhere in between, and IARPA-style tournaments are a method of figuring out roughly where we are in that conceptual space at the moment, with the caveat that things can always change suddenly.

I recall Daniel Kahneman having said on a number of occasions that when he's talking to people in large organizations, private or public sector, he challenges the seriousness of their commitment to improving judgment and choice. The challenge takes the following form: would you be willing to devote 1 percent of your annual budget to efforts to improve judgment and choice? And to the best of my knowledge, I don't think he's had any takers yet.

One of the things I've discovered in my work on assessing the accuracy of probability judgment is that there is much more eagerness in participating in these exercises among people who are younger and lower in status in organizations than there is among people who are older and higher in status in organizations. It doesn't require great psychological insight to understand this. You have a lot more to lose if you're senior and well established and your judgment is revealed to be far less well calibrated than that of people who are far junior to you.

Level-playing-field forecasting exercises are radically meritocratic. They put everybody on the same playing field. Tom Friedman no longer has an advantage over an unknown columnist, or for that matter, an unknown graduate student. If Tom Friedman's subjective probability estimate for how things are going in the Middle East is less accurate than that of the graduate student at Berkeley, the forecasting tournament just cranks through the numbers and that's what you discover.

These are potentially radically status-destabilizing interventions. They have the potential to destabilize status relationships within government agencies. They have the potential to destabilize the status within the private sector. The primary claim that people higher in status organizations have to holding their positions is cognitive in nature. They know better. They know things that the people below them don't know. And insofar as forecasting exercises are probative and give us insight into who knows what about what, they are, again, status destabilizing.

From a sociological point of view, it's a minor miracle that this forecasting tournament is even occurring. Government agencies are not supposed to sponsor exercises that have the potential to embarrass them. It would be embarrassing if it turns out that thousands of amateurs working on relatively small budgets are able to outperform professionals within a multibillion-dollar bureaucracy. That would be destabilizing. If it turns out that junior analysts within that multibillion-dollar bureaucracy can perform better than people high up in the bureaucracy, that would be destabilizing. If it turns out that the CEO is not nearly as good as people two or three tiers down in perceiving strategic threats to the business, that's destabilizing.

Things that bring transparency to judgment are dangerous to your status. You can make a case for this happening in medicine, for example. Insofar as evidence-based medicine protocols become increasingly influential, doctors are going to rely more and more on the algorithms—otherwise they're not going to get their bills paid. If they're not following the algorithms, it's not going to be reimbursable. When the health-care system started to approach 20 to 25 percent of the GDP, very powerful economic actors started pushing back and demanding accountability for medical judgment.

The long and the short of the story is that it's very hard for professionals and executives to maintain their status if they can't maintain a certain mystique about their judgment. If they lose that mystique about their judgment, that's profoundly threatening. My inner sociologist says to me that when a good idea comes up against entrenched interests, the good idea typically fails. But this is going to be a hard thing to suppress. Level-playing-field forecasting tournaments are going to spread. They're going to proliferate. They're fun. They're informative. They're useful in both the private and public sectors. There's going to be a movement in that direction. How it all sorts out is interesting. To what extent is it going to destabilize the existing pundit hierarchy? To what extent is it going to destabilize who the big shots are within organizations?

The Intelligence Advance Research Projects Agency about two years ago committed to supporting five university-based research teams and funded their efforts to recruit forecasters, set up websites for eliciting forecasts, hire statisticians for aggregating forecasts, and conduct a variety of experiments on factors that might either make forecasters more accurate or less accurate. For about a year and half we've been doing actual forecasting.

There are two aspects of this. There's a horse race aspect to it and there's a basic science aspect. The horse race aspect is, which team is more accurate? Which team is generating probability judgments that are closer to reality? What does it mean to generate a probability judgment closer to reality? If I say there is a .9 likelihood of Obama winning reelection and Nate Silver says there's a .8 likelihood of Obama reelection and Obama wins reelection, the person who said .9 is closer than the person who said .8. So that person deserves a better accuracy score. If someone said .2, they get a really bad accuracy score.

There are some statistical procedures that we use for a method of scoring probability judgment. It's called Brier scoring. Brier scoring is what we are using right now for assessing accuracy, but there are many other statistical techniques that can be applied. Our conclusions are robust across them. But the idea is to get people to make explicit probability judgments and score them against reality. Yet to make this work, you also have to pose questions that could be resolved in a clear-cut way. You can't say, "I think there could be a lot of instability in Afghanistan after NATO forces withdraw in 2014." That's not a good question.

The questions need to pass what some psychologists have called the "Clairvoyance Test." A good question would be one that I could take and hand to a genuine clairvoyant and ask, "What happened? What's the true answer?" A clairvoyant could look into the future and tell me about it. The clairvoyant wouldn't have to come back to me and say, "What exactly do you mean by 'could'?" or "What exactly do you mean by 'increased violence' or 'increased instability' or 'unrest'?" or whatever the other vague phrases are. We would have to translate them into something testable.

An important part of this forecasting tournament is moving from interesting issues to testable propositions, and this is an area where we discover very quickly where people don't think the way that Karl Popper thought they should think—like falsificationists. We don't naturally look for evidence that could falsify our hunches, and passing the Clairvoyance Test requires doing that.

If you think that the Eurozone is going to collapse—if you think it was a really bad idea to put into common currency economies at very different levels of competitiveness, like Greece and Germany (that was a fundamentally unsound macroeconomic thing to do and the Eurozone is doomed)—that's a nice example of an emphatic but untestable hedgehog kind of statement. It may be true, but it's not very useful for our forecasting tournament.

To make a forecasting tournament work we have to translate that hedgehog-like hunch into a testable proposition like, will Greece leave the Eurozone or formally withdraw from the Eurozone? Or will Portugal? You need to translate the abstract interesting issue into testable propositions and then you need to get lots of thoughtful people to make probability judgments in response to those testable proposition questions. You need to do that over, and over, and over again.

Hedgehogs are more likely to embrace fast and frugal heuristics that are in the spirit of blink. If you have a hedgehog-like framework, you're more likely to think that people who have mastered that framework should be able to diagnose situations quite quickly and reach conclusions quite confidently. Those things tend to co-vary with each other.

For example, if you have a generic theory of world politics known as "realism" and you believe that when there's a dominant power being threatened by a rising power, say the United States being threatened by China, it's inevitable that those two countries will come to blows in some fashion—if you believe that, then blink will come more naturally to you as a forecasting strategy.

If you're a fox and you believe there's some truth to the generalization that rising powers and hegemons tend to come into conflict with each other, but there are lots of other factors in play in the current geopolitical environment that make it less likely that China and the United States will come into conflict—that doesn't allow blink anymore, does it? It leads to "on the one hand, and on the other" patterns of reasoning—and you've got to strike some kind of integrative resolution of the conflicting arguments.

In the IARPA tournament, we're looking at a number of strategies for improving prediction. Some of them are focused on the individual psychological level of analysis. Can we train people in certain principles of probabilistic reasoning that will allow them to become more accurate? The answer is, to some degree we can. Can we put them together in collaborative teams that will bring out more careful self-critical analysis? To some degree we can. Those are interventions at the individual level of analysis.

Then the question is, you've got a lot of interesting predictions at the individual level—what are you going to do with them? How are you going to combine them to make a formal prediction in the forecasting tournament? It's probably a bad idea to take your best forecaster and submit that person's forecasts. You probably want something a little more statistically stable than that.

That carries us over into the wisdom-of-the-crowd argument—

the famous Francis Galton country fair episode in which the average of 500 or 600 fairgoers make a prediction about the weight of an ox. I forget the exact numbers, but let's say the estimated average prediction was 1,100. The individual predictions were anywhere from 300 to 14,000. When we trim outliers and average, it came to 1,103, and the true answer was 1,102. The average was more accurate than all of the individuals from whom the average was derived. I haven't got all the details right there, but that's a stylized representation of the aggregation argument.

There is some truth to that in the IARPA tournament. That simple averaging of the individual forecasters helps. But you can take it further: you can go beyond individual averaging and you can move to more complex weighted averaging kinds of formulas of the sort—for example, that Nate Silver and various other polimetricians were using in the 2012 election. But we're not aggregating polls anymore; we're aggregating individual forecasters in sneaky and mysterious ways. Computers are an important part of this story.

In *Moneyball* algorithms destabilized the status hierarchy. You remember in the movie, there was this nerdy kid amid the seasoned older baseball scouts, and the nerdy kid was more accurate than the seasoned baseball scouts. It created a lot of friction there.

This is a recurring theme in the psychological literature—the tension between human-based forecasting and machine or algorithm-based forecasting. It goes back to 1954. Paul Meehl wrote on clinical versus actuarial prediction in which clinical psychologists and psychiatrists' predictions were being compared to various algorithms. Over the last 58 years there have been hundreds of studies done comparing human-based prediction to algorithm- or machine-based prediction, and the track

Philip Tetlock

record doesn't look good for people. People just keep getting their butts kicked over and over again.

We don't have geopolitical algorithms that we're comparing our forecasters to, but we're turning our forecasters into algorithms, and those algorithms are outperforming the individual forecasters by substantial margins. There's another thing you can do, though, and it's more the wave of the future. And that is, you can go beyond human versus machine or human versus algorithm comparison or Kasparov versus Deep Blue (the famous chess competition) and ask, how well could Kasparov play chess if Deep Blue were advising him? What would the quality of chess be there? Would Kasparov and Deep Blue have an FIDE chess rating of 3,500, as opposed to Kasparov's rating of, say, 2,800 and the machine's rating of, say, 2,900? That is a new and interesting frontier for work, and it's one we're experimenting with.

In our tournament, we've skimmed off the very best forecasters in the first year, the top 2 percent. We call them "super forecasters." They're working together in five teams of 12 each and they're doing very impressive work. We're experimentally manipulating their access to the algorithms as well. They get to see what the algorithms look like, as well as their own predictions. The question is—do they do better when they know what the algorithms are, or do they do worse?

There are different schools of thought in psychology about this, and I have some very respected colleagues who disagree with me on it. My initial hunch was that they might be able to do better. Some very respected colleagues believe that they're probably going to do worse.

The most amazing thing about this tournament is that it exists because it is so potentially status-destabilizing. Another amazing and wonderful thing about this tournament is how many really smart, thoughtful people are willing to volunteer essentially enormous amounts of time to make this successful. We offer them a token honorarium. We're paying them right now $150 or $250 a year for their participation. The ones who are really taking it seriously—it's way less than minimum wage. And they're some very thoughtful professionals who are participating in this. Some political scientists I know have had some disparaging things to say about the people who might participate in something like this, and one phrase that comes to mind is "unemployed news junkies." I don't think that's a fair characterization of our forecasters. Certainly the most actively engaged of our forecasters are really pretty awesome. They're very skillful at finding information, synthesizing it, and applying it, and then updating the response to new information. And they're very rapid updaters.

There is a saying that's very relevant to this whole thing, which is that "life only makes sense looking backward, but it has to be lived going forward." My life has just been a quirky path-dependent meander. I wound up doing this because I was recruited in a fluky way to a National Research Council Committee on American Soviet Relations in 1983 and 1984. The Cold War was at its height. I was, by far, the most junior member of the committee. It was fluky that I became engaged in this activity, but I was prepared for it in some ways. I'd had a long-standing interest in marrying political science and psychology. Psychology is not just a natural biological science. It's a social science, and a great deal of psychology is shaped by social context.

Philip Tetlock

3
Smart Heuristics

Gerd Gigerenzer

Psychologist; Director of the Center for Adaptive Behavior and Cognition, Max Planck Institute for Human Development, Berlin; author, *Calculated Risk*, *Gut Feelings*, and *Risk Savvy*

INTRODUCTION by John Brockman

"Isn't more information always better?" asks Gerd Gigerenzer. "Why else would bestsellers on how to make good decisions tell us to consider all pieces of information, weigh them carefully, and compute the optimal choice, preferably with the aid of a fancy statistical software package? In economics, Nobel prizes are regularly awarded for work that assumes that people make decisions as if they had perfect information and could compute the optimal solution for the problem at hand. But how do real people make good decisions under the usual conditions of little time and scarce information? Consider how players catch a ball—in baseball, cricket, or soccer. It may seem that they would have to solve complex differential equations in their heads to predict the trajectory of the ball. In fact, players use a simple heuristic. When a ball comes in high, the player fixates on the ball and starts running. The heuristic is to adjust the running speed so that the angle of gaze remains constant—that is, the angle between the eye and the ball. The player can ignore all the information necessary to compute the trajectory, such as the ball's initial velocity, distance, and angle, and just focus on one piece of information, the angle of gaze."

Gigerenzer provides an alternative to the view of the mind as a cognitive optimizer, and also to its mirror image, the mind as a cognitive miser. The fact that people ignore information has been often mistaken as a form of irrationality, and shelves are filled with books that explain how people routinely commit cognitive fallacies. In seven years of research, he and his research team at the Center for Adaptive Behavior and Cognition at the Max Planck Institute for Human Development in Berlin have worked out what he believes is a viable alternative: the study of fast and frugal decision making, that is, the study of smart heuristics people actually use to make good decisions. In order to make good decisions in an uncertain world, one sometimes has to ignore information. The art is knowing what one doesn't have to know.

Gigerenzer's work is of importance to people interested in how the human mind actually solves problems. In this regard his work is influential to psychologists, economists, philosophers, and animal biologists, among others. It is also of interest to people who design smart systems to solve problems; he provides illustrations on how one can construct fast and frugal strategies for coronary care unit decisions, personnel selection, and stock picking.

"My work will, I hope, change the way people think about human rationality," he says. "Human rationality cannot be understood, I argue, by the ideals of omniscience and optimization. In an uncertain world, there is no optimal solution known for most interesting and urgent problems. When human behavior fails to meet these Olympian expectations, many psychologists conclude that the mind is doomed to irrationality. These are the two dominant views today, and neither extreme of hyper-rationality or irrationality captures the essence of human reasoning. My aim is not so much to criticize the status quo, but rather to provide a viable alternative."

At the beginning of the 20th century the father of modern science fiction, Herbert George Wells, said in his writings on politics, "If we want to have an educated citizenship in a modern technological society, we need to teach them three things: reading, writing, and statistical thinking." At the beginning of the 21st century, how far have we gotten with this program? In our society, we teach most citizens reading and writing from the time they are children, but not statistical thinking. John Alan Paulos has called this phenomenon innumeracy.

There are many stories documenting this problem. For instance, there was the weather forecaster who announced on American TV that if the probability that it will rain on Saturday is 50 percent and the probability that it will rain on Sunday is 50 percent, the probability that it will rain over the weekend is 100 percent. In another recent case reported by *New Scientist*, an inspector in the Food and Drug Administration visited a restaurant in Salt Lake City famous for its quiches made from four fresh eggs. She told the owner that according to FDA research every fourth egg has salmonella bacteria, so the restaurant should only use three eggs in a quiche. We can laugh about these examples because we easily understand the mistakes involved, but there are more serious issues. When it comes to medical and legal issues, we need exactly the kind of education that H. G. Wells was asking for, and we haven't gotten it.

What interests me is the question of how humans learn to live with uncertainty. Before the scientific revolution, determinism was a strong ideal. Religion brought about a denial of uncertainty, and many people knew that their kin or their race was exactly the one that God had favored. They also thought they were entitled to get rid of competing ideas and the people who propagated them. How does a society change from this condition into one in which we understand that there is this fundamental uncertainty? How do we avoid the illusion of certainty to produce the understanding that everything, whether it be a medical test or deciding on the best cure for a particular kind of cancer, has a fundamental element of uncertainty?

For instance, I've worked with physicians and physician-patient associations to try to teach the acceptance of uncertainty and the reasonable way to deal with it. Take HIV testing as an example. Brochures published by the Illinois Department of Health say that testing positive for HIV means that you have the virus. Thus, if you are an average person who is not in a particular risk group but test positive for HIV, this might lead you to choose to commit suicide, or move to California, or do something else quite drastic. But AIDS information in many countries is running on the illusion of certainty. The actual situation is rather like this: If you have about 10,000 people who are in no risk group, one of them will have the virus, and will test positive with practical certainty. Among the other 9,999, another one will test positive, but it's a false positive. In this case we have two who test positive, although only one of them actually has the virus. Knowing about these very simple things can prevent serious disasters, of which there is unfortunately a record.

Still, medical societies, individual doctors, and individual patients either produce the illusion of certainty or want it. Everyone knows Benjamin Franklin's adage that there is nothing certain in this world except death and taxes, but the doctors I interviewed tell me something different. They say, "If I would tell my patients what we don't know, they would get very nervous, so it's better not to tell them." Thus, this is one important area in which there is a need to get people—including individual doctors or lawyers in court—to be mature citizens and to help them understand and communicate risks.

Representation of information is important. In the case of many so-called cognitive illusions, the problem results from difficulties that arise from getting along with probabilities. The problem largely disappears the moment you give the person the information in natural frequencies. You basically put the mind back in a situation where it's much easier to understand these probabilities. We can prove that natural frequencies can facilitate actual computations, and have known for a long time that representations— whether they be probabilities, frequencies, or odds—have an impact on the human mind. There are very few theories about how this works.

I'll give you a couple examples relating to medical care. In the United States and many European countries, women who are 40 years old are told to participate in mammography screening. Say that a woman takes her first mammogram and it comes out positive. She might ask the physician, "What does that mean? Do I have breast cancer? Or are my chances of having it 99 percent, 95 percent, or 90 percent, or only 50 percent? What do we know at this point?" I have put the same question to radiologists who have done mammography screening for 20 or 25 years, including chiefs of departments. A third said they would tell this woman that, given a positive mammogram, her chance of having breast cancer is 90 percent.

However, what happens when they get additional relevant information? The chance that a woman in this age group has cancer is roughly 1 percent. If a woman has breast cancer, the probability that she will test positive on a mammogram is 90 percent. If a woman does not have breast cancer the probability that she nevertheless tests positive is some 9 percent. In technical terms you have a base rate of 1 percent, a sensitivity or hit rate of 90 percent, and a false-positive rate of about 9 percent. So how do you answer this woman who's just tested positive for cancer? As I just said, about a third of the physicians thinks it's 90 percent, another third thinks the answer should be something between 50 percent and 80 percent, and another third thinks the answer is between 1 percent and 10 percent. Again, these are professionals with many years of experience. It's hard to imagine a larger variability in physicians' judgments—between 1 percent and 90 percent—and if patients knew about this variability, they would not be very happy. This situation is typical of what we know from laboratory experiments: namely, that when people encounter probabilities—which are technically conditional probabilities—their minds are clouded when they try to make an inference.

What we do is to teach these physicians tools that change the representation so that they can see through the problem. We don't send them to a statistics course, since they wouldn't have the time to go in the first place, and most likely they wouldn't understand it because they would be taught probabilities again. But how can we help them to understand the situation?

Let's change the representation using natural frequencies, as if the physician would have observed these patients him- or herself. One can communicate the same information in the following, much more simple way. Think about 100 women. One of them has breast cancer. This was the 1 percent. She likely tests positive; that's the 90 percent. Out of 99 who do not have breast cancer another 9 or 10 will test positive. So we have one in 9 or 10 who tests positive. How many of them actually has cancer? One out of ten. That's not 90 percent, that's not 50 percent, that's one out of ten.

Here we have a method that enables physicians to see through the fog just by changing the representation, turning their innumeracy into insight. Many of these physicians have carried this innumeracy around for decades and have tried to hide it. When we interview them, they obviously admit it, saying, "I don't know what to do with these numbers. I always confuse these things." Here we have a chance to use very simple tools to help those pa-

Gerd Gigerenzer

tients and physicians understand what the risks are, and which enable them to have a reasonable reaction to what to do. If you take the perspective of a patient—that this test means that there is a 90 percent chance you have cancer—you can imagine what emotions set in, emotions that do not help her to reason the right way. But informing her that only one out of ten women who tests positive actually has cancer would help her to have a cooler attitude and to make more reasonable decisions.

Prostate cancer is another disease for which we have good data. In the United States and European countries doctors advise men aged 40 to 50 to take a PSA test. This is a prostate cancer test that is very simple, requiring just a bit of blood, so many people do it. The interesting thing is that most of the men I've talked to have no idea of the benefits and costs of this test. It's an example of decision making based on trusting your doctor or on rumors. But interestingly, if you read about the test on the Internet in independent medical societies like Cochran.com, or read the reports of various physicians' agencies who give recommendations for screening, then you find out that the benefits and costs of prostate cancer screening are roughly the following: Mortality reduction is the usual goal of medical testing, yet there's no proof that prostate cancer screening reduces mortality. On the other hand there is proof that if we distinguish between people who do not have prostate cancer and those who do, there is a good likelihood that it will do harm. The test produces a number of false positives. If you do it often enough there's a good chance of getting a high level on the test, a so-called positive result, even though you don't have cancer. It's like a car alarm that goes off all the time.

For those who actually have cancer, surgery can result in incontinence or impotence, which are serious consequences that stay with you for the rest of your life. For that reason, the U.S. Preventive Services task force says very clearly in a report that men should not participate in PSA screening because there is no proof in mortality reduction, only likely harm.

It is very puzzling that in a country where a 12-year-old knows baseball statistics, adults don't know the simplest statistics about tests, diseases, and the consequences that may cause them serious damage. Why is this? One reason, of course, is that the cost-benefit computations for doctors are not the same as for patients. One cannot simply accuse doctors of knowing things or not caring about patients, but a doctor has to face the possibility that if he or she doesn't advise someone to participate in the PSA test and that person gets prostate cancer, then the patient may turn up at his doorstep with a lawyer. The second thing is that doctors are members of a community with professional pride, and, for many of them, not detecting a cancer is something they don't want to have on their records. Third, there are groups of doctors who have very clear financial incentives to perform certain procedures. A good doctor would explain this to a patient but leave the decision to the patient. Many patients don't see this situation in which doctors find themselves, but most doctors will recommend the test.

But who knows? Autopsy studies show that one out of three or one out of four men who die a natural death has prostate cancer. Everyone has some cancer cells. If everyone underwent PSA testing and cancer were detected, then these poor guys would spend the last years or decades of their lives living with severe bodily injury. These are very simple facts.

Thus, dealing with probabilities also relates to the issue of understanding the psychology of how we make rational decisions. According to decision theory, rational decisions are made according to the so-called expected utility calculus, or some variant thereof. In economics, for instance, the idea is that if you make an important

Gerd Gigerenzer

decision—whom to marry or what stock to buy, for example—you look at all the consequences of each decision, attach a probability to these consequences, attach a value, and sum them up, choosing the optimal, highest expected value or expected utility. This theory, which is very widespread, maintains that people behave in this way when they make their decisions. The problem is that we know from experimental studies that people don't behave this way.

There is a nice story that illustrates the whole conflict: A famous decision theorist who once taught at Columbia got an offer from a rival university and was struggling with the question of whether to stay where he was or accept the new post. His friend, a philosopher, took him aside and said, "What's the problem? Just do what you write about and what you teach your students. Maximize your expected utility." The decision theorist, exasperated, responded, "Come on, get serious!"

Decisions can often be modeled by what I call fast and frugal heuristics. Sometimes they're faster, and sometimes they're more frugal. Deciding which of two jobs to take, for instance, may involve consequences that are incommensurate from the point of view of the person making the decision. The new job may give you more money and prestige, but it might leave your children in tears, since they don't want to move for fear that they would lose their friends. Some economists may believe that you can bring everything in the same common denominator, but others can't do this. A person could end up making a decision for one dominant reason.

We make decisions based on a bounded rationality, not the unbounded rationality of the decision maker modeled after an omniscient god. But bounded rationality is also not of one kind. There is a group of economists, for example, who look at the bounds or constraints in the environment that affect how a decision is made. This study is called "optimization under constraints," and many

Nobel prizes have been awarded in this area. Using the concept of bounded rationality from this perspective, you realize that an organism has neither unlimited resources nor unlimited time. So one asks, given these constraints, what's the optimal solution?

There's a second group, which doesn't look at bounds in the environment but at bounds in the mind. These include many psychologists and behavioral economists who find that people often take in only limited information, and sometimes make decisions based on just one or two criteria. But these colleagues don't analyze the environmental influences on the task. They think that for a priori reasons people make bad choices because of a bias, an error, or a fallacy. They look at constraints in the mind.

Neither of these concepts takes advantage of what the human mind takes advantage of: that the bounds in the mind are not unrelated to the bounds in the environment. The bounds get together. Herbert Simon developed a wonderful analogy based on a pair of scissors, where one blade is cognition and the other is the structure of the environment, or the task. You only understand how human behavior functions if you look at both sides.

Evolutionary thinking gives us a useful framework for asking some interesting questions that are not often posed. For instance, when I look at a certain heuristic—like when people make a decision based on one good reason while ignoring all others—I must ask in what environmental structures that heuristic works, and where it does not work. This is a question about ecological rationale, about the adaptation of heuristics, and it is very different from what we see in the study of cognitive illusions in social psychology and of judgment decision-making, where any kind of behavior that suggests that people ignore information, or just use one or two pieces of information, is coded as a bias. That approach is nonecological: that is, it doesn't relate the mind to its environment.

Gerd Gigerenzer

An important future direction in cognitive science is to understand that human minds are embedded in an environment. This is not the usual way that many psychologists, and of course many economists, think about it. There are many psychological theories about what's in the mind, and there may be all kinds of computations and motives in the mind, but there's very little ecological thinking about what certain cognitive strategies or emotions do for us, and what problems they solve. One of the visions I have is to understand not only how cognitive heuristics work, and in which environments it is smart to use them, but also what role emotions play in our judgment. We have gone through a kind of liberation in the last years. There are many books, by Antonio Damasio and others, that make a general claim that emotions are important for cognitive functions, and are not just there to interrupt, distract, or mislead you. Actually, emotions can do certain things that cognitive strategies can't do, but we have very little understanding of exactly how that works.

To give a simple example, imagine *Homo economicus* in mate search, trying to find a woman to marry. According to standard theory *Homo economicus* would have to find out all the possible options and all the possible consequences of marrying each one of them. He would also look at the probabilities of various consequences of marrying each of them—whether the woman would still talk to him after they're married, whether she'd take care of their children, whatever is important to him—and the utilities of each of these. *Homo economicus* would have to do tons of research to avoid just coming up with subjective probabilities, and after many years of research he'd probably find out that his final choice had already married another person who didn't do these computations, and actually just fell in love with her.

Herbert Simon's idea of satisfying solves that problem. A satisfier, searching for a mate, would have an aspiration level. Once this aspiration is met, as long as it is not too high, he will find the partner and the problem is solved. But satisfying is also a purely cognitive mechanism. After you make your choice you might see someone come around the corner who looks better, and there's nothing to prevent you from dropping your wife or your husband and going off with the next one.

Here we see one function of emotions. Love, whether it be romantic love or love for our children, helps most of us to create a commitment necessary to make us stay with and take care of our spouses and families. Emotions can perform functions that are similar to those that cognitive building blocks of heuristics perform. Disgust, for example, keeps you from eating lots of things and makes food choice much simpler, and other emotions do similar things. Still, we have very little understanding of how decision theory links with the theory of emotion, and how we develop a good vocabulary of building blocks necessary for making decisions. This is one direction in which it is important to investigate in the future.

Another simple example of how heuristics are useful can be seen in the following thought experiment: Assume you want to study how players catch balls that come in from a high angle—like in baseball, cricket, or soccer—because you want to build a robot that can catch them. The traditional approach, which is much like optimization under constraints, would be to try to give your robot the complete representation of its environment and the most expensive computation machinery you can afford. You might feed your robot a family of parabolas because thrown balls have parabolic trajectories, with the idea that the robot needs to find the right parabola in order to catch the ball. Or you feed him measurement instruments that can measure the initial distance, the initial velocity, and the initial angle the ball was thrown or kicked. You're

Gerd Gigerenzer

still not done because in the real world balls are not flying parabolas, so you need instruments that can measure the direction and the speed of the wind at each point of the ball's flight to calculate its final trajectory and its spin. It's a very hard problem, but this is one way to look at it.

A very different way to approach this is to ask if there is a heuristic that a player could actually use to solve this problem without making any of these calculations, or only very few. Experimental studies have shown that actual players use a quite simple heuristic that I call the gaze heuristic. When a ball comes in high, a player starts running and fixates his eyes on the ball. The heuristic is that you adjust your running speed so that the angle of the gaze, the angle between the eye and the ball, remains constant. If you make the angle constant the ball will come down to you and it will catch you, or at least it will hit you. This heuristic only pays attention to one variable, the angle of gaze, and can ignore all the other causal, relevant variables and achieve the same goal much faster, more frugally, and with less chances for error.

This illustrates that we can do the science of calculation by looking always at what the mind does—the heuristics and the structures of environments—and how minds change the structures of environments. In this case the relationship between the ball and one's self is turned into a simple linear relationship on which the player acts. This is an example of a smart heuristic, which is part of the adaptive toolbox that has evolved in humans. Many of these heuristics are also present in animals. For instance, a recent study showed that when dogs catch frisbees they use the same gaze heuristic.

Heuristics are also useful in very important practical ways relating to economics. To illustrate I'll give you a short story about our research on a heuristic concerning the stock market. One very smart and simple heuristic is called the recognition heuristic. Here is a demonstration: Which of the following two cities has more inhabitants—Hanover or Bielefeld? I pick these two German cities assuming that you don't know very much about Germany. Most people will think it's Hanover because they have never heard of Bielefeld, and they're right. However, if I pose the same question to Germans, they are insecure and don't know which to choose. They've heard of both of them and try to recall information. The same thing can be done in reverse. We have done studies with Daniel Gray Goldstein in which we ask Americans which city has more inhabitants—San Diego or San Antonio. About two-thirds of my former undergraduates at the University of Chicago got the right answer: San Diego. Then we asked German students—who know much less about San Diego and many of whom had never even heard of San Antonio—the same question. What proportion of the German students do you think got the answer right? In our study, 100 percent. They hadn't heard of San Antonio, so they picked San Diego. This is an interesting case of a smart heuristic, where people with less knowledge can do better than people with more. The reason this works is because in the real world there is a correlation between name recognition and things like populations. You have heard of a city because there is something happening there. It's not an indicator of certainty, but it's a good stimulus.

In my group at the Max Planck Institute for Human Development I work alongside a spectrum of researchers, several of whom are economists, who work on the same topics but ask a different kind of question. They say, "That's all fine that you can demonstrate that you can get away with less knowledge, but can the recognition heuristic make money?" In order to answer this question we did a large study with the American and German stock markets, involving both laypeople and students of business and finance in both countries. We went to downtown Chicago and interviewed several hundred pedestrians. We gave them a list of stocks and asked them one question: Have you ever heard of this stock? Yes or no? Then we took the ten percent of the stocks that had the highest recognition, which were all stocks in the Standard & Poor's Index, put them in the portfolio, and let them go for half a year. As a control, we did the same thing with the same American pedestrians with German stocks. In this case they had heard of very few of them. As a third control we had German pedestrians in downtown Munich perform the same recognition ratings with German and American stocks. The question in this experiment is not how much money the portfolio makes, but whether it makes more money than some standards, of which we had four. One consisted of randomly picked stocks, which is a tough standard. A second one contained the least-recognized stocks, which is, according to the theory, an important standard, and shouldn't do as well. In the third we had blue chip funds, like Fidelity II. And in the last we had the market—the Dow and its German equivalent. We let this run for six months, and after six months the portfolios containing the highest recognized stocks by ordinary people outperformed the randomly picked stocks, the low-recognition stocks, and in six out of eight cases the market and the mutual funds.

Although this was an interesting study, one should of course be cautious, because unlike in other experimental and real-world studies, we have a variable and very random environment. But what this study at least showed is that the recognition of ordinary citizens can actually beat out the performance of the market and other important criteria. The empirical evidence, of course— the background—is consumer behavior. In many situations when people in a supermarket choose between products, they go with the item with name recognition. Advertising by companies like

Benetton exploits the use of the recognition heuristic. They give us no information about the product, but only increase name recognition. It has been a very successful strategy for the firm.

Of course the reaction to this study, which is published in our book, *Simple Heuristics That Make Us Work*, has split the experts in two camps. One group said this can't be true, that it's all wrong, or it could never be replicated. Among them were financial advisers, who certainly didn't like the results. Another group of people said, "This is no surprise. I knew it all along. The stock market's all rumor, recognition, and psychology." Meanwhile, we have replicated these studies several times and found the same advantage of recognition—in bull and bear markets—and also found that recognition among those who knew less did best of all in our studies.

I would like to share these ideas with many others, to use psychological research, and to use what we know about how to facilitate people's understanding of uncertainties to help promote this old dream about getting an educated citizenship that can deal with uncertainties, rather than denying their existence. Understanding the mind as a tool that tries to live in an uncertain world is an important challenge.

Gerd Gigerenzer

4

Affective Forecasting . . . Or . . . The Big Wombassa: What You Think You're Going to Get, and What You Don't Get, When You Get What You Want

Daniel Gilbert

Professor of Psychology, Harvard University; author, *Stumbling on Happiness*.

INTRODUCTION by John Brockman

In 1968, I was sitting at the back of a seedy Sunset Strip nightclub in Hollywood having a few drinks with a friend, an actor, who, in the blink of an eye, had unexpectedly become an overnight sensation. An actor's actor, highly regarded by his peers, he was now in the middle of the actor's dream fantasy . . . on the covers of national newsmagazines, women, ac-claim, and . . . work.

Or was it an actor's nightmare?

"The big wombassa," he said quietly.

"What?" I asked. "The big what?"

"It's 'the big wombassa,' Johnny, what you think you're going to get, and what you don't get, when you get what you want."

The Big Wombassa.

Indeed, over the years I've thought a lot about this highly intuitive for-mulation, but it wasn't until I met Harvard psychologist Daniel Gilbert that I found out that syndromes such as "The Big Wombassa" had become a legitimate subject of scientific inquiry, and at no less a place than Harvard's

Social Cognition and Emotion Lab, which is bringing scientific rigor to the study of subjective experiences such as satisfaction and happiness.

Gilbert is well-known for his work on what he and his long-time collaborator, Tim Wilson of the University of Virginia, call affective forecasting, which is "the ability to predict one's hedonic reactions to future events."

He points out that "many economists believe that affective forecasting errors are impediments to rational action and hence should be eliminated—just as we would all agree that illiteracy or innumeracy are bad things that deserve to be eradicated. But cognitive errors may be more like optical illusions than like illiteracy. The human visual system is susceptible to a variety of optical illusions, but if someone offered to surgically restructure your eyes and your visual cortex so that parallel lines no longer appeared to converge on the horizon, you should run as far and fast as possible."

Gilbert approaches these issues as a scientist, not a clinician. He is "interested in learning how people can become better affective forecasters, but not because I believe that people should become better affective forecasters. My job as a scientist is to find and explain these errors and illusions, and it is up to each individual to decide how they want to use our findings."

Economic decisions are inherently affective forecasts. Economists believe that people engage in economic transactions in order to "maximize their utility." Now, for psychologists the word *utility* isn't particularly meaningful unless you are talking about gas and electricity. Psychologists argue that utility is actually a stand-in for something like happiness or satisfaction—some subjective, hedonic state of the decision maker. That sounds a bit squishy to modern economists, who often confuse utility with wealth, but how could it be otherwise?

Daniel Gilbert

People engage in economic transactions in order to get things that they believe will provide them with positive emotional experiences, and wealth is nothing more than an "experience credit" that can be used to attain those experiences in the future. So rational economic behavior requires that we look into the future and figure out what will provide that experience and what won't. As it turns out, people make systematic errors when they do this, which is why their economic decisions are so often suboptimal.

The problem lies in how we imagine our future hedonic states. We are the only animals that can peer deeply into our futures—the only animal that can travel mentally through time, preview a variety of futures, and choose the one that will bring us the greatest pleasure and/or the least pain. This is a remarkable adaptation—which, incidentally, is directly tied to the evolution of the frontal lobe—because it means that we can learn from mistakes before we make them. We don't have to actually have gallbladder surgery or lounge around on a Caribbean beach to know that one of these is better than another. We may do this better than any other animal, but our research suggests that we don't do it perfectly. Our ability to simulate the future and to forecast our hedonic reactions to it is seriously flawed, and people are rarely as happy or unhappy as they expect to be.

What kinds of errors and mistakes do people make? The first thing to note is that psychologists who study errors of judgment are only interested in systematic errors. There's a difference between an error and a systematic error. If you're standing in front of a dart board and you're trying to hit the bull's-eye, you are bound to miss sometimes, but your errors will be randomly distributed around the middle of the dartboard. The mere fact that you can't hit the bull's-eye every time is not particularly interesting or unusual, and the mere fact that people are inaccurate in predicting their hedonic reactions to future events is not interesting or unusual either. But if every time you missed the bull's-eye you made a particular kind of error—for example, if all of your misses were twenty degrees to the left—then something interesting and unusual might be happening, and we might start to wonder what it was.

Perhaps you have a visual deficit, perhaps the dart is badly weighted, perhaps there is a strong air current in the room. Systematic errors beg for scientific explanations, and as it turns out, the errors that people make when they try to predict their emotional futures are quite systematic. Specifically, people tend to overestimate the impact of future events. That is, they predict that future events will have a more intense and more enduring hedonic impact than they actually do. We call this the impact bias.

Let me give you a couple of real-world examples of this bias. We've done dozens of studies in both the laboratory and the field, and the general strategy of the research is really very simple: We ask people to predict how they will feel minutes, days, weeks, months, or even years after some future event occurs, and then we measure how they actually do feel after that event occurs. If the two numbers differ systematically, then we have one of those interesting and unusual systematic errors I mentioned.

We've seen the impact bias in just about every context we've studied. For example, we've studied numerous elections over the last few years, and voters invariably predict that if their candidate wins they're going to be happy for months, and if their candidate loses they'll be unhappy for months. In fact, their happiness is barely influenced by electoral outcomes.

We see the same pattern when we look at the dissolution of romantic relationships. People predict that they will be very unhappy for a very long time after a romantic relationship dissolves, but the fact is that they are usually back to their baseline in a

Daniel Gilbert

relatively short time—a much shorter time than they predicted. Professors expect to be happier for years after getting tenure than after being denied tenure, but the two groups are equally happy in a brief time. Please understand that I'm not saying that these events had no impact. Of course promotions make us feel good and divorces make us feel bad! What I'm saying is that whatever impact these events have, it is demonstrably smaller and less enduring than the impact the people who experienced them expected them to have.

Now, notice something about these events: they are remarkably ordinary. We aren't asking people to tell us how they'll feel after a Martian invasion. Most voters have voted and won before, most lovers have loved and lost before. For the most part, the events we study are events that people have experienced many times in their lives—events about which they should be quite expert—which makes their inaccuracy all the more curious and all the more interesting.

The question, then, is not whether there is an impact bias, because that has been amply demonstrated both by our lab and by others. The question is, why? Why are we such strangers to ourselves? There are a couple of different answers to this question. Most robust phenomena in nature are multiply determined, which is to say that when something happens all the time there are probably a lot of independent mechanisms making it happen. That's what we've found with the impact bias. Let me tell you about a few of the mechanisms that seem to give rise to the impact bias.

First, people have a tremendous talent for changing their views of events so that they can feel better about them. We're not immediately delighted when our wife runs away with another guy, but in fairly short order most of us start to realize that "she was never really right for me" or that "we didn't have that much in common."

Our friends snicker and say that we are rationalizing—as if these conclusions were wrong simply because they are comforting. In fact, rationalization doesn't necessarily mean self-delusion. These conclusions may actually have been right all along, and rationalization may be the process of discovering what was always true but previously unacknowledged. But it really doesn't matter from my perspective whether these conclusions are objectively true or not.

What matters is that human beings are exceptionally good at discovering them when it is convenient for them to do so. Shakespeare wrote, "'Tis nothing either good or bad, but thinking makes it so," and in fact, thinking is a remarkable tool that allows us to change our views of the world in order to change our emotional reactions to the world. Once we discover how wrong our wife was for us, her departure is transformed from a trauma to a blessing.

Now, it's not big news that people are good at this. What is news is that people don't know they're good at this. Rationalization is largely an unconscious process. We don't wake up in the morning and say, "Today I'm going to fool myself." Rather, soon after a bad event occurs, unconscious processes are activated, and these processes begin to generate different ways of construing the event. Thoughts such as "Maybe I was never really in love" seem to come to mind all by themselves, and we feel like the passive recipients of a reasonable suggestion. Because we don't consciously experience the cognitive processes that are creating these new ways of thinking about the event, we don't realize they will occur in the future.

One of the reasons that we think bad things will make us feel bad for a long time is that we don't realize that we have this defensive system—something like a psychological immune system, if you will. If I were to ask you to predict how healthy you would be if you encountered a cold germ and you didn't know that you had a physical immune system, you'd expect to get very sick and perhaps even die.

Similarly, when people predict how they're going to feel in the face of adversity, not knowing they have a psychological immune system leads them to expect more intense and enduring dissatisfaction than they will actually experience. We have several studies demonstrating this point. For example, if you ask subjects in an experiment to predict how they will feel a few minutes after getting negative feedback about their personalities from a clinician or a computer, they expect to feel awful—and they expect to feel equally awful in both cases.

But when you actually give them that feedback, they feel slightly disappointed but not awful. Moreover, they feel much less disappointed when the feedback came from a computer than from a clinician. Why? Because it is much easier to rationalize feedback from a computer than a clinician. After all, what does a machine know? What's interesting is that subjects don't realize in prospect that they will do this. Results such as these suggest that people just aren't looking forward to their opportunities for rationalization when they predict their future happiness.

Consider another mechanism that causes the impact bias. I spend a lot of time asking people to imagine how they would feel a year after their child died (as you can imagine, this makes me very popular at parties). Everybody gives the same answer, of course, which is some form of "I would be totally devastated." Then I ask them what they did to come to that conclusion, and they'll almost always report that they had a horrifying mental image of being at a funeral at which their child is being buried, or of standing in the child's room looking at an empty crib, etc. These horrifying images serve as the basis for their predictions, which, as it turns out, are wrong. The clinical literature suggests that people who lose a child are not usually "thoroughly devastated" a year later. The event has lasting repercussions, of course, but what is remarkable about the people who experience it is just how well they usually do. As your grandmother said, life goes on.

So why do people mispredict their reactions to tragedies like this one?

A mental image captures one moment of a single event. But one's happiness a year after the event is influenced by much more than the event itself. A lot happens in a year—there are birthday parties, school plays, promotions, love-making, dental appointments, hot fudge sundaes, and so on. These things aren't nearly as important as the tragedy, of course, but they are real, there are a lot of them, and together they have an impact that forecasters tend not to consider.

When we're trying to predict how happy we will be in a future that contains Event X, we tend to focus on Event X and forget about all the other events that also populate that future—events that tend to dilute the hedonic impact of Event X. In a sense, we are slaves to the focus of our own attention. For example, in one study we asked college students to predict how happy or unhappy they would be a few days after their home team won or lost a football game, and they expected the game to have a large impact on their hedonic state. But when we simply asked them to name a dozen other things that would happen in those days before they made their predictions, the game had far less impact on their predictions. In other words, once they thought about how well-populated the future was, they realized that the game was just one of many sources of happiness and that its impact would be diluted by others.

When you study errors such as these, it is only natural to wonder

Daniel Gilbert

how they might be avoided, and people are always asking me if I would like to develop programs to improve people's affective forecasting accuracy. Before we rush out to develop such programs, we should ask whether the impact bias is something we want to live without. Errors in human judgment are logical violations—if you say you'll feel 7 on a 1-to-10 scale and you actually end up feeling 5, then you've made a mistake. But is that mistake a bad thing?

The fact is that errors can have adaptive value. For example, perhaps it is important for organisms to believe they would be thoroughly devastated by the loss of their offspring, and the fact that this isn't actually true is beside the point. What may matter is that the organism thinks it is true and acts accordingly. Perhaps the best way to think of an error in judgment is like a mosquito in an ecosystem. You see the darn pest and your first inclination is to ask, how do get we rid of these? So you spray DDT and you kill all the mosquitoes and then you find out that the mosquitoes were at the bottom of a food chain and the fish ate the mosquitoes, and the frogs ate the fish, and the bears ate the frogs, and now the entire ecology is devastated. Similarly, errors in human judgment may be playing important roles that scientists don't see.

Many economists believe that affective forecasting errors are impediments to rational action and hence should be eliminated—just as we would all agree that illiteracy and innumeracy are bad things that deserve to be eradicated. But cognitive errors may be more like optical illusions than they are like illiteracy. The human visual system is susceptible to a variety of optical illusions, but if someone offers to surgically restructure your eyes and your visual cortex so that parallel lines no longer appeared to converge on the horizon, you should run as far and fast as possible.

I'm interested in learning how people can become better affective forecasters, but not because I believe that people should become better affective forecasters. My job as a scientist is to find and explain these errors and illusions, and it is up to each individual to decide how they want to use our findings.

With that said, our research does suggest that there is a simple antidote to affective forecasting errors. Consider this. There are two ways to make a prediction about how you're going to feel in the future. The first is to close your eyes and imagine that future—to simulate it in your own mind and preview your own hedonic reaction. That's the kind of affective forecasting we've studied extensively, and what we now know is that the process of projecting oneself into the future is fraught with error. But there's a second way to make these kinds of forecasts, namely, to find somebody who's already experiencing that future and observe how they actually feel.

If you were trying to decide whether you should take job X or job Y, you might try to imagine yourself in each of them, but you might instead observe people who have job X and job Y and simply see how happy they are. What we've discovered is that (a) when people do this, they make extremely accurate affective forecasts, and (b) no one does this unless you force them to!

Try this thought experiment: You're going to go on a vacation to a tropical island. It's offered at a very good price, and you have to decide whether you're willing to pay. You are offered one of two pieces of information to help you make your decision. Either you can have a brochure about the hotel and the recreational activities on the island, or you can find out how much a randomly selected traveler who recently spent time there liked his or her experience. Which would you prefer? In studies we've done that are modeled on this thought experiment, roughly 100 percent of the people prefer the kind of information contained in the brochure. After all, who the hell wants to hear from some random guy when they can look at the brochure and judge for themselves?

Nonetheless, if you actually give people one of these two pieces of information, they more accurately predict their own happiness when they see the random traveler's report then when they see the brochure. Why? Because the brochure enables you to simulate what the island might be like and how much you'd enjoy it, but as I've mentioned, these sorts of predictions are susceptible to a wide variety of errors.

On the other hand, another person's report enables you to avoid these errors because it allows you to base your predictions on real experience rather than imaginary experience. If another person liked the island, the odds are that you will like it too. There's a delicious irony here, which is that the information we need to predict how we'll feel in the future is usually right in front of us in the form of other people. But because individuals believe so much in their own uniqueness—because we think we're so psychologically different from others—we refuse to use the information that's right before our eyes.

If you want to be a better affective forecaster, then, you would do well to base your forecasts on the actual experiences of real people who've been in the situations you are only imagining. The more similar to you the person is, the more informative their experience will be, of course. But what's amazing is that even the experience of a randomly selected person provides a better basis for forecasting than does your own imagination.

If you actually looked at the correlates of happiness across the human population, you learn a few important things. First of all, wealth is a poor predictor of happiness. It's not a useless predictor, but it is quite limited. The first $40,000 or so buys you almost all of the happiness you can get from wealth. The difference between earning nothing and earning $20,000 is enormous—that's the difference between having shelter and food and being homeless and hungry.

But economists have shown us that after basic needs are met, there isn't much "marginal utility" to increased wealth. In other words, the difference between a guy who makes $15,000 and a guy who makes $40,000 is much bigger than the difference between the guy who makes $100,000 and the guy who makes $1,000,000. Psychologists, philosophers, and religious leaders are a little too quick to say that money can't buy happiness, and that really betrays a failure to understand what it's like to live in the streets with an empty stomach. Money makes a big difference to people who have none.

On the other hand, once basic needs are met, further wealth doesn't seem to predict further happiness. So the relationship between money and happiness is complicated, and definitely not linear. If it were linear, then billionaires would be a thousand times happier than millionaires, who would be a hundred times happier than professors. That clearly isn't the case.

On the other hand, social relationships are a powerful predictor of happiness—much more so than money is. Happy people have extensive social networks and good relationships with the people in those networks. What's interesting to me is that while money is weakly and complexly correlated with happiness, and social relationships are strongly and simply correlated with happiness, most of us spend most of our time trying to be happy by pursuing wealth. Why?

Individuals and societies don't have the same fundamental need. Individuals want to be happy, and societies want individuals to consume. Most of us don't feel personally responsible for stoking our country's economic engine; we feel personally responsible

Daniel Gilbert

for increasing our own well-being. These different goals present a real dilemma, and society cunningly solves it by teaching us that consumption will bring us happiness.

Society convinces us that what's good for the economy is good for us too. This message is delivered to us by every magazine, television, newspaper, and billboard, at every bus stop, grocery store, and airport. It finds us in our cars, it's made its way onto our clothing. Happiness, we learn, is just around the corner and it requires that we consume just one more thing. And then just one thing more after that. So we do. We find out that the happiness of consumption is thin and fleeting, and rather than thinking to ourselves, "Gosh, that promise of happiness-by-consumption was a lie," we instead think, "Gosh, I must not have consumed enough and I probably need just one small upgrade to my stereo, car, wardrobe, or wife, and then I'll be happy."

We live in the shadow of a great lie, and by the time we figure out that it is a lie we are closing in on death and have become irrelevant consumers, and a new generation of young and relevant consumers takes our place in the great chain of shopping.

Do I make all these affective forecasting errors myself? You bet I do. Because of the research I do, I occasionally glimpse life from the experimenter's point of view, but most of the time I'm just another one of life's subjects and I do all the same things that everyone does. I make the same mistakes the other subjects make, and if there is any difference between us it is that I am dimly aware of my mistakes as I make them.

But awareness isn't enough to stop me. Affective forecasting errors are a bit like perceptual illusions in this respect. Someone shows you a neat illusion and you say, "Wow, it looks like the black rectangle is floating above the white one even though it's really not." But that awareness doesn't make the illusion go away. Similarly, you can know at an intellectual level that an affective forecast is wrong, but that in and of itself doesn't change the fact that it feels so damn right. For example, my girlfriend is a consultant who has to live in different cities five days a week, and I am absolutely convinced that if she would just find a job in Cambridge and be home with me at night, I would be deliriously happy forevermore. I am as convinced as anyone that my big wombassa is just around the corner.

Daniel Gilbert

5

Adventures in Behavioral Neurology— Or—What Neurology Can Tell Us About Human Nature

Vilayanur Ramachandran

Director of the Center for Brain and Cognition and Distinguished Professor with the Psychology Department and the Neurosciences Program, University of California–San Diego; Adjunct Professor of Biology, Salk Institute; coauthor, *Phantoms in the Brain*; author, *The Tell-Tale Brain*.

I'm interested in all aspects of the human mind, including aspects of the mind that have been regarded as ineffable or mysterious. The way I approach these problems is to look at patients who have sustained injury to a small region in the brain, a discipline called behavioral neurology or cognitive neuroscience these days.

Let me tell you about the problem confronting us. The brain is a 1.5-kilogram mass of jelly, the consistency of tofu, and you can hold it in the palm of your hand, yet it can contemplate the vastness of space and time, the meaning of infinity, and the meaning of existence. It can ask questions about who am I, where do I come from, questions about love and beauty, aesthetics and art, and all these questions arise from this lump of jelly. It is truly the greatest of mysteries. The question is, how does it come about?

When you look at the structure of the brain, it's made up of neurons. Of course, everybody knows that these days. There are 100 billion of these nerve cells. Each of these cells makes about

1,000 to 10,000 contacts with other neurons. From this information people have calculated that the number of possible brain states, of permutations and combinations of brain activity, exceeds the number of elementary particles in the universe.

The question is, how do you go about studying this organ? There are various ways of doing it. These days brain imaging is very popular. You make the person perform some task, engage in conversation or think about love, for that matter, or something like that, or imagine the color red. What part of the brain lights up? That gives you some confidence in saying that that region of the brain is involved in mediating that function. I'm sort of simplifying it, but something along those lines. Then there is recording from single cells, where you put an electrode through the brain, eavesdrop on the activity of individual neurons, and find out what the neuron is responsive to in the external world. There are dozens of such approaches, and our approach is behavioral neurology combined with brain imaging.

Behavioral neurology has a long history going back about 150 years, a venerable tradition going back to Charcot. Even Freud was a behavioral neurologist. We usually think of him as a psychologist, but he was also a neurologist. In fact, he began his career as a neurologist, comparable in stature with Charcot, Hughling Jackson, and Kurt Goldstein. What they did was look at patients with sustained injury to a very small region of the brain—and this is what we do as well in our lab. What you get is not a blunting of all your mental capacities or across-the-board reduction of your mental ability. What you get often is a highly selective loss of one specific function, other functions being preserved relatively intact. This gives you some confidence in saying that that region of the brain is specialized in dealing with that function.

It doesn't have to be a lesion; it can be a genetic change. One of the phenomena that we've studied, for example, is synesthesia, the merging of the senses (which I'll talk about in a minute) where there has been a genetic glitch. It runs in families in whom some gene or genes cause people to hear colors and taste sounds. They've got their senses muddled up. We've been studying this phenomenon.

In general, we look at curious phenomena, syndromes that have been known for ages, maybe 100 years, 50 years, that people have brushed under the carpet because they're regarded as anomalies, to use Thomas Kuhn's phrase. What do you make of somebody who says, "I see five as red, six as blue, seven as green, F sharp as indigo?" It doesn't make any sense, and when you see this in science, the tendency among most scientists, most of my colleagues at any rate, is to brush it under the carpet and pretend it doesn't exist, deny it. What we do is to go and rescue these phenomena from oblivion, studying them intensively in the laboratory. Nine out of ten times it's a wild-goose chase, but every now and then you hit the jackpot and you discover something really interesting and important. This is what happened with synesthesia. Another example, which maybe I'll begin with, is one most people have heard of—our work on phantom limbs and mirrors, which I'll touch on in a minute.

One of the peculiar syndromes that we have studied recently is called apotemnophilia. It's in fact so uncommon that many neurologists and many psychiatrists have not heard of it. It's in a sense a converse of phantom limbs. In a phantom-limb patient, an arm is amputated but the patient continues to vividly feel the presence of that arm. We call it a phantom limb.

In apotemnophilia you are dealing with a perfectly healthy, normal individual, not mentally disturbed in any way, not psychotic, not emotionally disturbed, often holding a job, and has a family. We saw a patient recently who was a prominent dean of an engineering school, and soon after he retired he came out and said he wants his left arm amputated above the elbow. Here's a perfectly normal guy who has been living a normal life in society and interacting with people. He's never told anybody that he harbored this secret desire—intense desire—to have his arm amputated ever since early childhood, and he never came out and told people about it for fear that they might think he was crazy. He came to see us recently and we tried to figure out what was going on in his brain. And by the way, this disorder is not rare. There are websites devoted to it. About one-third of them go on to actually get it amputated—not in this country because it's not legal, but they go to Mexico or somewhere else and get it amputated.

So here is something staring you in the face, an extraordinary syndrome, utterly mysterious, where a person wants his normal limb removed. Why does this happen? There are all kinds of crazy theories about it, including Freudian theories. One theory asserts, for example, that it's an attention-seeking behavior. This chap wants attention, so he asks you to remove his arm. It doesn't make any sense. Why does he not want his nose removed or ear removed or something less drastic? Why an arm? It seems a little bit too drastic for seeking attention.

The second thing that struck us is that the guy would often take a felt pen and draw a very precise irregular line around his arm or leg and say, "I want it removed exactly that way. I don't want you removing too little of it or too much of it. It would feel wrong. I want you to amputate it exactly on that line." And you could test him after a year—it is the same wiggly line which he couldn't have memorized, and this suggests already that this is something physiological, and not something psychological that he is making up.

Vilayanur Ramachandran

Another theory that is even more absurd (found in some papers, and again, it's also a Freudian theory) is that the guy wants a big stump because it resembles a giant penis. Sort of wish fulfillment. This again is ridiculous, complete nonsense, of course. The question is, why does it actually happen? What we were struck by was that there are certain syndromes where the patient has a right hemisphere stroke, in the right parietal cortex. The patient then starts denying that the left arm belongs to him. He says, "Doctor, this arm"—he'll often point to it with his right arm—"this arm belongs to my mother." Here's a person who is perfectly coherent, intelligent, can discuss politics with you, can discuss mathematics with you, play chess with you, asserting that his left arm doesn't belong to him.

This is different from apotemnophilia. In apotemnophilia the patient says, "This arm is mine, but I don't want it. I want it removed." But there are similarities, there's an overlap, so we suggested that maybe there's something wrong with his body image in the right hemisphere which alienates the left arm, or the right arm, for that matter, from the rest of the person's body, and the sense of alienation leads to the person saying, "I don't want it. Have it removed."

More specifically, messages from the arm and the skin throughout the body, in fact, go to the parietal lobe to a structure, the postcentral gyrus. There's a big furrow, or cleft, right down the middle of the brain called the central sulcus. Just behind that sulcus there is a vertical, narrow strip of cortex where there's a complete map of your body's surface. Every point of your body's surface is represented in a specific point on the cortex and there's a complete map called the Penfield Map. That's where touch sensations and, behind it, joint sensations and muscle sensations are all represented in this somatosensory map.

The first thing we—Paul McGeoch, Dave Brang, and I—did was an MEG recording (MEG is a functional brain imaging technique) to map out the body of these people. Normal people have a complete point-by-point map on the surface of this strip of cortex. We said, well, maybe this guy has a hole in that region corresponding to the arm he wants removed because it feels alien. But what we found is that there are no holes. It's a completely normal map, so we were disappointed. Then what we found was that there's another region behind it, called the superior parietal lobule. This region actually constructs your body image. When I close my eyes I have a vivid sense of my different body parts. Some parts are more vivid than others and this comes mainly from joint and muscle sense, partly from my vestibular sense—saying that I'm standing erect, that my head is not tilted—and partly that when I open my eyes I confirm it with vision. So there's a convergence of signals from vision, touch, proprioception, vestibular sense, vision—all of that—helping you construct a vivid internal picture of your own body, called the body image. That gets partial input from the map I was telling you about, namely the touch map, the map for joints and muscle sense. It also gets input from hearing. It gets input from vestibular sense. It gets input from vision. All of it is converging to create a body image. That map, we found, does not have the representation of the arm that the patient wants to get rid of (you don't see this in every patient; in some the malfunction may be in the zones that the body image map subsequently projects to).

So our hypothesis was the signals arrive in S1 (touch) and S2 (joint and muscle sense). All the sensory signals arrive here and they're all normal and they're received in the brain, but when the signal gets sent to the body image center in the superior parietal lobule in the right hemisphere there is no place in the brain to receive that signal and, therefore, this creates a tremendous clash and discrepancy, and the brain abhors discrepancies. The discrepancy signal is then sent to the amygdala, the limbic structure produces an aversion to the arm, and the patient says, "I want the arm removed. It feels intrusive"—he uses words like "intrusive" over "present"—"so I want it removed."

Here's a bizarre psychological syndrome: the person wants his arm removed. Discard the Freudian idea that he wants a big, giant stump or that he wants attention and things like that, and you come up with a precise circuit diagram of what's going on.

We tested this because it is not enough to come up with a theory. How do we test it? It turns out that if I poke anybody with a needle, that pain sensation goes to the sensory pain region in the brain, probably in the thalamus and cortex, and then it goes to the amygdala. The amygdala alerts you to the pain and you say, "Ow." Right? And then it goes down to the anterior cingulate that feels the agony of the pain. There's a cascade of events. There's a sensation of pain and then the agony of pain, and then messages go down the autonomic nervous system and make you start sweating, preparing for action, fleeing, fighting, or whatever the required action is. So if you poke somebody with a needle, this whole cascade of events is set in motion. You can measure the skin resistance, which measures the sweating—you start sweating when poked.

Now, when I poked him with a little pencil above the line where he wanted his arm to be amputated, nothing happened. Just a little gentle pencil prick. It's not a painful stimulus. Not much happened. There's no galvanic skin response, there's no arousal. But if you touch him below the line where he wants the amputation, there's a huge, big galvanic skin response. You can actually measure the aversion physiologically, not just rely on the subject to report. There's no way you can fake the galvanic skin response. It's the basis of all lie detector tests.

Then, of course, we went straight to the brain and said, let's map it out. And as I said, we found S1 was normal; if you go to S2 that's normal. If you go to the superior parietal lobule, where the body image is constructed, to some extent the inferior parietal lobule, right parietal, let's say, where the body image is, there is no arm representation in that center. That's what we found. If you touch the arm there's no activity there. If you touch above the line of amputation or touch the normal arm, the activity is completely normal. So that region of the brain is abnormal, but we also speculate that in the regions of the brain in the frontal lobes and insula/amygdala to which that SPL/IPL projects, there could be an interruption of signals. In either event there's a physiological reason why this happens. This is giving you insight into how the normal brain constructs a body image.

At this point I should add a note of caution that unlike our work on phantoms and synesthesia, which has been confirmed on dozens of subjects—both by us and others—this work on apotemnophilia is very preliminary; we need more subjects. And that's not a legal disclaimer—it's the truth. Let's wait and see. Then the question is, can you treat the guy? And we're working on that. We don't know how. In the case of phantom limb, there are ways of treating phantom pain.

We don't deliberately go after these odd phenomena. Somebody phones me and says, "I have this curious phenomenon, Dr. Ramachandran. Can you solve the problem?" Ninety percent of the time we can't, but every now and then we discover something amazing, as I said.

So what we do is, we look at the patient, and the first question is, why does this patient have this syndrome? Why this peculiar behavior? What's going on in the brain? Can you explain it? First of all, show that he's not crazy. Show that it's a real, authentic syndrome. There are bogus syndromes and I will talk about that if you like. They're not real. But given that it's a genuine syndrome and you prove that it's authentic, which is often very difficult to do, then, having done that, you say, "What are the precise brain mechanisms that give rise to these curious symptoms?" So the first two questions are: is it real? And if so, what causes it? The third question is, who cares? What does it matter? Each of these three questions needs to be answered if you want to make progress, if you want to draw people's attention to the phenomenon that you're studying.

So let's take synesthesia. First thing we did was to show that these people are not crazy. They're really seeing colors when they see numbers. They're not just making it up. The second thing we did was to ask, what are the brain mechanisms, what's going on in the brain. Ed Hubbard, Boynton, and I have shown that when they see noncolored numbers the color area in the brain lights up. So what? Why should I care? We've shown that this quirky phenomenon has broad implications for understanding creativity and metaphor. Well, we haven't actually shown it, but we've speculated on that possibility. So in each case what we do is we rescue this phenomenon, this anomaly, from being brushed under the carpet, find out what the mechanism is in the brain, and talk about its deeper implications for all of us, for normal behavior.

Now, we seek odd syndromes to try to explain the symptoms in neural terms and hopefully shed light on aspects of human nature that have remained ineffable for the longest time. But sometimes you come across a syndrome where you cannot quite know for sure if this is a legit syndrome or not, even though you can find it in the bible of clinical psychologists called *DSM, Diagnostic and*

Statistical Manual, which is the official book for clinicians. If they can label you, give your syndrome a name, they can charge you, charge an insurance company, so there has been a tendency to multiply syndromes.

There's one called, by the way, Chronic Underachievement Syndrome, which in my day used to be called stupidity. It actually has a name and it's officially recognized. Then there is a syndrome called De Clérambault's Syndrome. De Clérambault's Syndrome refers to, believe it or not, a young woman developing an obsession with a much older, famous, eminent, rich guy and develops the delusion that that guy is madly in love with her but is in denial about it. This is actually found in a textbook of psychiatry, and I think it's complete nonsense. Ironically, there's no name for the converse of the syndrome, where an aging male develops a delusion that this young hottie is madly in love with him but is in denial about it. Surely it's much more common, and yet it doesn't have a name. Right?

You have to have a nose for anomalies, nose for the right kinds of syndromes to pursue. I'll give you examples, and I think two will suffice. Let's return to synesthesia, which I have been excited about studying for the last three or four years. It's not really a neurological syndrome, but it is an anomaly of sorts. Francis Galton described it in the 19th century, the great Victorian polymath who was a first cousin of Charles Darwin. Galton noticed that some people in the population were otherwise quite normal, except they had one quirk: every time they saw a number, the number would evoke a specific color. So five would be tinged red, six would be green, seven would be indigo, eight would be blue, and nine would be orange, or something like that. Also, sometimes in the same subjects or in other people, each tone would evoke a color. So F sharp would be blue, C sharp would be green, so on and so forth.

Galton also noticed that this runs in families and may have a genetic basis and published this, I think it was in *Nature*, in the 19th century. Since then there have been dozens and dozens of case reports of people experiencing this, but it was regarded as a curiosity mainly and is also thought to be very rare, estimates ranging from one in 1,000 to one in 10,000.

I became very intrigued by this phenomenon and I said, "Why would somebody see five as red?" Now, as I said, it's been known for a long time, for more than 100 years, and people ignored it because it didn't fit the big framework of science. What do you make of someone who says F sharp is blue and C sharp is green? It doesn't make any sense. Or five is green or five is red? And I tend to get intrigued by these phenomena. I said, "Well, what's going on in their brain?" The first thing we wanted to show was that this was a legitimate phenomenon, that these people are not making it up.

There are several theories of synesthesia. One theory is that they are crazy. Maybe, but let's set that aside for a minute. One of the things we learn in medicine is that when a patient is trying to tell you something and you think he's crazy, it often means that you're not smart enough to figure it out. Sometimes he's crazy, but usually it means you're not smart enough to figure it out, so look carefully, talk to the patient.

In the case of synesthesia, another odd aspect of it is that it's much more common among artists, poets, novelists, and other creative people. In fact, seven or eight times more common. This is controversial, but the strong evidence is that this is true. Now why would that be the case? I mean, one of my students has shown this to be the case, Ed Hubbard—why would this happen? There are several little mini-mysteries here about synesthesia. Why would it run in families? Why would they say numbers are colors or tones are colors, for that matter? Why would it be more common in artists, poets, and novelists? So on and so forth. So it's a medical mystery worthy of Sherlock Holmes, waiting to be solved.

The first thing we want to show is that these people are not crazy. By the way, another common theory is that they are on drugs like LSD—acid junkies or potheads. Sure enough it's more common among people who are high on acid, but that makes it even more intriguing. Why would some drugs produce this merging of the senses, this peculiar phenomenon of numbers evoking colors?

Another theory is that they're being metaphorical, as when you and I say, "It is the east and Juliet is the sun." Or we just say, "This is a loud tie." The tie isn't loud. It doesn't make any sound. Why do you say it's a loud tie? Or cheese. Cheddar cheese is sharp. Now, sharp is a tactile adjective, a sharp nail or something. Why do you use a tactile adjective to describe a taste, a gustatory sensation? You say, well, it's a metaphor. That's circular. Why do you want to use a tactile metaphor for a taste sensation?

Explaining synesthesia as just a metaphor doesn't explain anything because it's trying to explain one mystery with another mystery, and that doesn't work in science. Another example of a metaphor would be, "It is the east and Juliet is the sun." You don't say, "Juliet is the sun." Does that mean she's a glowing ball of fire? No, you don't say that. You say, "She is warm like the sun. She is radiant like the sun." "Is nurturing like the sun," is a celestial body like the sun (a pun rather than a metaphor), "is the center of my solar system," and so on. The brain forms the right links. Synesthesia, by the way, is a completely arbitrary link between five and red. It's not a metaphor in that sense, so I was uncomfortable with the idea, but I thought there might be something to it. But we'll come back to that later as we go along. About a decade ago, by the way, I proposed there may be unconscious synesthetic propensities in all of us, which has now been amply confirmed in many studies, including a recent one from Oxford.

Another theory is that they're remembering childhood memories. Maybe they played with refrigerator magnets and five was red, and six was blue, and seven was green, and for some reason they're stuck with these memories. Well, this again begs the question of why you and I have played with magnets but we don't have synesthesia, presumably. Most of us don't. We found, by the way, the phenomenon of synesthesia is quite common. You see it in one in 50 people. It's not one in 1,000 or one in 10,000. People often don't come out and say that they do because they're worried you might think they're crazy.

So the childhood memories thing doesn't work because, as I said, why would it run in families? Another reason for not believing it—you would have to say the same magnets were being passed from generation to generation, and it doesn't make any sense. Metaphor? Maybe they are being metaphorical in some sense. Maybe it's related to metaphors. They're crazy? That's not a real argument. They're on drugs—no, that doesn't work either.

The first thing we wanted to show is they're not crazy. They aren't making this up. We generated a computerized display made up of fives, lots of scattered fives on the screen, and among those fives were scattered some twos. When you look at a two and a five, a five is a mirror of a two in a sense, in terms of its shape. So you have a bunch of outline drawings of fives. Scattered among them are some twos forming a shape. The twos cluster to form a triangle or a square or a circle like your Ishihara color test in traffic, when you're going through a traffic school eye exam for color vision. It's similar to that.

A normal person looking at it, the nonsynesthete looking at this, says, "Oh, fives? You mean there are twos in here embedded? Let me see. Oh, there's a two there. There's a two. Okay. Oh, there's a two there. There's another one there." They take 20 or 30 seconds to find the hidden shape. A synesthete who sees five as red and two as green instantly sees a green circle or a green square or a hidden green shape pop out from the background. He's much faster in detecting the circle or the square than you and I are. If he's crazy, how come he's better at it than us? Secondly, if you ask him what he sees, he says, "I see a green triangle. I see a green square." Phenomenologically, perceptually, he literally sees the green square or the triangle or the rectangle. What this suggests is that it's a sensory experience, not a memory association, at least in some synesthetes. Jamie Ward has recently replicated our findings.

It turns out there is a heterogeneity of synesthetes. There are some synesthetes, whom we will call lower synesthetes, in whom the color is actually perceptually evoked and the numbers seem tinged with color—red, green, blue, yellow, chartreuse, or indigo. But there are also more conceptual synesthetes where it does seem to be more like a memory association. We were focusing on the perceptual synesthetes, sensory synesthetes, because they are easier to study scientifically.

First, we've shown they're not crazy, it's a real phenomenon. (Remember, I had three steps: First, to show it's real. Second, what are the brain mechanisms? Third, what are the broader implications? Why should I care?) We've solved the first problem, which is to show that it's a genuine phenomenon.

The second question is, what causes it? Well, Ed Hubbard and I were looking at brain atlases and we were struck by the fact that there's a structure called the fusiform gyrus in the brain buried inside the folds of the temporal lobes. This structure, the fusiform gyrus, it turns out, is where the color area of the brain is, V4, which was discovered by Semir Zeki. Right next to it, almost touching it, is the number area of the brain. It represents the visual representations of numbers. The two areas are almost touching each other. We said, what's the likelihood that the most common type of synesthesia is the number-color synesthesia, and the number region and color region are adjacent to each other in the brain? This seems unlikely to be a coincidence. Then we said, maybe there's an accidental cross-wiring between these two regions of the brain.

How do you prove that? We did brain imaging. You take a normal person and do an FMR, functional magnetic resonance imaging, or MEG, and show them numbers. Only the number area in the fusiform gyrus will light up. Show colored numbers to a normal person, and V4, the color area, and the number area will both light up. If you show a synesthete a black-and-white number, both the number area and color area light up, thereby directly proving that there's a cross-activation going on.

Now, I should add that three out of four groups has shown it to be the case. There's one group who claim they don't see the activation. There's always uncertainty in brain imaging—inherently there is some noise—so it has not entirely been nailed down, but I think it's very likely that we are on the right track. Romke Rowlte in Amsterdam has studied this and she has also shown that there is an actual increase in white matter, actual fibers connecting V4 (color) and the number area within the fusiform gyrus, so that's about as good as it gets.

Now why would this happen? Why do these people have this cross-wiring? That's the next question. A clue first comes from observations made by Francis Galton, and has been confirmed since then: it runs in families, it may have a genetic basis. So we said that if you take the infant brain, a fetal brain, there's a tremendous redundancy of connections. Everything is connected to everything. It's a crude approximation, but it's almost true. Then what happens is that there are pruning genes, which prune away the excess connections between adjacent brain regions (or even separated brain regions that were densely connected). This creates a characteristic modularity of the adult brain. Now, if something goes wrong with the pruning gene—if pruning fails to occur in adjacent brain regions, like the color and number areas remaining connected even in the adult—if the gene is selectively expressed in the fusiform gyrus through transcription factors, for example, if it's expressed in the fusiform gyrus, then you'll get a number/color synesthete. Every time your guy sees a number, because of the cross-wiring, the color neurons are going to be activated. Every time he sees a number he sees a color.

In our early papers we noted that such cross-activation could also be based on transmitters that cause disinhibition; probably both things are going on. Voila, you explain number/color synesthesia. You started with a gene. The gene has not been cloned yet, but people are trying. You explain what's going on in the brain, why these people have these quirky visual experiences of seeing colored numbers.

The last question—why should I care? The answer comes from the observation, the claim, that synesthesia is seven or eight times more common among artists, poets, and novelists. Artists like Kandinsky, for example. Why would this be the case? What do artists, poets, and novelists have in common? They're all very good at metaphor and analogy, seeing hidden links that most of us lesser mortals have difficulty in seeing. So when Shakespeare said, "It is the east and Juliet is the sun," as I have said before, you don't say, "Juliet is the sun," meaning that she's a glowing ball of fire. You make the right links, you say she's celestial like the sun. You make any number of links. She's the center of my solar system like the sun is the center of the solar system. She's radiant like the sun. She's warm like the sun. She's nurturing like the sun. Shakespeare was very good at picking these metaphors, which have multiple layers of metaphors and resonance.

What has this got to do with synesthesia? What's going on in a metaphor? You're linking seemingly unrelated concepts and ideas, right? If the same synesthesia gene, instead of being expressed selectively in the fusiform gyrus and producing this quirky phenomenon of number/color synesthesia, if it were to be expressed throughout the cortex, throughout the brain, it's going to create a higher propensity and higher opportunity to link seemingly unrelated ideas and concepts in far-flung brain regions. If we think of ideas and concepts as also located in specific brain regions, occupying specific brain regions, and if you have these long-range connections, then it permits greater opportunity for linking seemingly unrelated concepts. Hence the basis of creativity and metaphors. Hence the eight-times-higher incidence of synesthesia among artists, poets, and novelists.

In other words, what I'm getting at is that an evolutionary biologist could ask the question, what use is this gene? It's seen in one in 50 people. It's fairly common, not rare. Why is it conserved in evolution? If there's a gene in evolution that's useless—it's completely useless to see five as red and six as green—it would have been eliminated from the gene pool eons ago, 10,000 years, 20,000 years ago. Clearly, this gene has been around and has been conserved. Now why? Why is this gene still around if it's completely useless?

Well, one possibility is that it confers some outliers in the population with the ability to link seemingly unrelated ideas, making them artistic, more creative. But when I give these talks, people often ask me why, if it's that good that that gene makes you artistic, creative, and metaphorical, why doesn't everybody have it? Well, it's a silly question because evolution takes time, and given another 20,000, 100,000, 50,000 years everybody will have this gene and we'll all be creative. But that's not the right answer. It may be a partial answer, but the real answer, I think, is that you don't want everyone being creative; we need engineers!

You can see what we've done here, as with apotemnophilia, but even more so with synesthesia. It started with this peculiar phenomenon where people see number as color or tones as color. Then from that we said, what is it, is it real? We showed that it was a real phenomenon using a number of perceptual psychological tests that can't be faked. From that we went to the brain anatomy, doing brain imaging, and showed what parts of the brain are involved. Then from that we were able to say there's a genetic basis. So from gene to neuroanatomy to perceptual phenomenology. Finally, all the way to metaphor, Shakespeare, and poetry. All from starting with this peculiar quirk called synesthesia. This is what we do with every one of our syndromes. Sometimes we're partially successful. Sometimes we're fully successful in doing this.

Given that our lab is well known for studying these odd quirks of human behavior and explaining what's going on in the brain, and showing there are broader implications for understanding human nature, human consciousness—these things that everybody is curious about—when people have something quirky, they come and phone me up. Or if a physician stumbles on a new case which he finds he can't explain, he or she will often phone me. Nine out of ten times I can't do anything about it, but every now and then I find out what's going on in the brain and discover it's something very intriguing and possibly important.

In the case of synesthesia, it was regarded mainly as a curiosity and an anomaly. People were just brushing it under the carpet and saying, "What do you make of somebody who says five is red? They're just making it up or they're crazy. That's why it's more common among artists, because we all know artists are a bit crazy anyway and they all want to draw attention to themselves." There are all kinds of silly theories floating around. In fact, some synesthetes were diagnosed as psychotic and diagnosed as having schizophrenia. They were told they were hallucinating colors. They were prescribed psychotropic drugs for the schizophrenia. Then they came to realize that they had this perfectly normal phenomenon, not normal (but not pathological either) phenomenon called synesthesia.

I think it's fair to say that we, and Jamie Ward and Julia Simner and a couple of other groups, came and revived interest in this field, brought it into the mainstream. So now there are about 20 or 30 books on synesthesia. Not 20 or 30, maybe a dozen books on synesthesia. On just this one topic. When I started this nearly ten years ago, nobody was interested in this topic. There was one book by a clinician, a neurologist named Cytowic, but he was a prophet talking in the wilderness. Nobody paid any attention. He didn't really do any definitive perceptual experiments on it. He just said, "Here's a phenomenon worthy of studying." We were the first to actually start doing experiments on the phenomenon—to show that it's authentic, show that it's perceptual, and then pin it down to brain anatomy.

The reason I was attracted to it was because I'm curious about neurological syndromes, given my background in clinical neurology, among other things. I began with being intrigued by phantom limbs. Patients would come into the clinic with an arm missing or a leg missing and continue to vividly feel the presence of that missing arm or leg. And again, it has been known for about 100 years and people thought of it as a curiosity, as a case study to be reported during grand rounds: "Here is a patient with phantom limb." Nobody knew what to make of it, and certainly there was no interest in mainstream neuroscience in phantom limbs.

What we found is quite intriguing: two or three things. One discovery goes back about 15 years. Let's say I'm the guy with the phantom limb, I've lost my left arm and you're the physician. You come and touch the left side of my face. I start feeling the stroking sensation in my phantom thumb! Even though you're stroking my face I feel it in my phantom. If you touch this region, it's my index finger, that's my pinky. There's a complete map of the missing hand on the face. Now, why would this be? Here again is the medical mystery.

I started thinking about, and drew inspiration from, animal studies that have shown that if you cut the nerves going from the arms to the spinal cord, what happens is a complete reorganization of the sensory map in the brain. What happens in this patient when you remove the arm?

You remember earlier, when I spoke in the context of apotemnophilia, I said there's a complete map of the body surface on the surface of the brain, the post central gyrus. There's a vertical furrow on the side of the brain. Behind that is the map. This map is systematic and point to point for the most part, but it turns out that the face area on the brain is right next to the hand area of the brain. It's a quirk and nobody knows why. The map is continuous and systematic, but oddly enough, the hand area is right next to the face area.

In an adult if you remove the arm, the hand area of the brain is now devoid of sensory input. It's hungry for new sensory input and it's not getting any sensory input. The sensory input from

Vilayanur Ramachandran

the face skin which normally only goes to the adjacent face area in the brain now invades the vacated territory corresponding to the missing hand and activates the hand cells in the brain. That, of course, misinforms higher centers in the brain that the hand is being stimulated. The patient then experiences the sensations as arising from the missing phantom limb. When you touch the face skin the message not only goes to the face area, but also activates the hand area in the brain. So you're getting cross-wiring between the hand area and the face area of the brain.

We did a ten-minute experiment to show this, and it challenged the doctrine in neurology that neural connections of the brain are laid down in the fetus and in early infancy, and once they've been laid down by the genome there's nothing you can do to change these connections in the adult brain. That's why if you have a lesion in the adult brain, say following a stroke, there's such little recovery of function, and why neurological syndromes are so difficult to treat, notoriously difficult to treat.

It was believed there was no plasticity in the brain connections. We showed in our experiment that, in fact, there's a tremendous scope for rewiring. So much so that over a two-centimeter distance in brain tissue in the cortex the face input has now invaded the hand territory of the brain. Then we did brain imaging and showed that this invasion had actually occurred, but we already knew this from the psychological experiment. So I guess my mind is primed to think about cross-connections in the brain.

Now that's an example of cross-connections caused by amputation depriving sensory input. In synesthesia, just like the face and the hand areas, the color and the number areas are right next to each other. I started thinking, well, maybe this is cross-wiring again. But in this case the cross-wiring is not due to deafferentation by removing the sensory input, but due to genes, given that it runs in families.

Typically what happens is, somebody phones me. It's usually a neurologist or psychiatrist. They say, "Here's a strange case of a patient with apotemnophilia or Capgras syndrome or some such syndrome. Can you take a look at the patient and tell us what you think?" The patient shows up in the laboratory or in my office and tells me what the problem is. I start thinking about it. You don't tell the patient ahead of time what your theory is because you don't want to cue them. Then you do various experiments on the patient and test your hypothesis about what's going on in the brain in terms of known anatomy and physiology of the brain, not some sort of mumbo jumbo psychological theory. Then you go and test the theory using brain imaging or doing simple psychological experiments.

Sometimes we're able to devise treatment for the patients. For example, in phantom limbs, two-thirds of the patients with phantom limbs experience excruciating pain. There's no known treatment. I should restate that: there are 20 known treatments; none of them work. So we started investigating it to develop a treatment for it. But sometimes even just explaining to the patient he's not crazy, telling him, "You've got a phantom limb. The reason for this is that something is going on in the brain," is a tremendous relief for him. Somebody has apotemnophilia and wants his arm removed. Telling him, "You're not crazy, it's not Freudian, it's a specific anatomical reason why you're experiencing this." Then you go to the next step and say you have this hypothesis about what's going on. You've tested it, you know what's going on in his brain. But can you actually help the patient?

In the case of phantom limbs we've done experiments to show that we can; but let me give you another example. There's a curious disorder that I've not talked about in the past. Candy McCabe in England is studying it, and we are also studying it, and it's called RSD, or reflex sympathetic dystrophy. It's another one of these disorders that is not rare. I'd say one out of 20 stroke patients has it. You also see it in patients who don't have stroke but have trivial injuries of the finger, like a metacarpal bone fracture that causes an injury with intense, excruciating pain.

It turns out there are two kinds of pain. We think of pain as one thing subjectively, but evolutionarily there are two kinds: there is acute pain and there's chronic pain. Acute pain occurs when you touch a flame or a hot kettle and you say, "Ouch," and you withdraw your hand. Chronic pain is when there's gangrene or a fracture, typically a fracture, and there's excruciating pain caused by the fracture, and your hand becomes immobilized—you don't withdraw it. What's the evolution? Even though they feel the same perceptually, evolutionarily they're very, very different.

The function of acute pain is to mobilize the hand and remove it from the source of tissue injury to protect the hand. Chronic pain is the exact opposite. When there's an injury to a metacarpal bone, your hand freezes up and gets "paralyzed" temporarily. It's excruciatingly painful. Any attempt to move it is painful, so you don't move the arm. In the case of acute pain you mobilize the arm rapidly. In the case of chronic pain you immobilize it. Why? Because moving it would cause further tissue injury. So it's a protective reflex—immobilization. And then, of course, as the injury heals you start moving your hand again and the pain goes away. That's a normal cause of events.

But in a certain proportion of patients, stroke patients, and in a certain proportion of patients with a tiny little fracture, even a little hairline fracture or ruptured ligament, the pain persists with a vengeance. Even after the injury is healed, even as the fracture is healed, the pain persists for weeks, months, years, sometimes for life, for decades. Not only does the pain persist, the hand gets swollen and paralyzed, the pain spreads over the entire hand. This from just an injury to one little bone, and it involves the entire hand, the entire forearm. There's swelling of the hand, swelling of the arm, warmth, inflammation—all of that takes place on the arm. Again, you're stuck with it, and there's no known treatment that works.

We started thinking about this and why this should be. Well, as I said, it's a reflex in mobilization and it's painful. Anytime you attempt to move the hand it causes excruciating pain, so the patient gives up and says, "I'm not going to move my hand." Sometimes what happens is you get stuck with this, and this we call "learned pain." Any attempt to move it, the signal that gets sent to the hand to move it, becomes associated with excruciating pain in your brain, so putting it crudely, you get a form of learned pain, a learned association between a motor command and the sensation of pain. The brain just gives up and the hand gets paralyzed. Any attempt to move it becomes excruciatingly painful.

How do you break the cycle? We said, "Let's use a mirror." So we put a mirror in the center of the table. This is similar to a mirror treatment for phantom pain and for stroke that we discovered over a decade ago. You put a mirror in the center of the table and the patient puts his painful dystrophic, swollen, immobilized, paralyzed arm on the left side of the mirror. The shiny side of the mirror is on the right side and the patient puts his right hand on the right side of the mirror, positions it so it mimics the posture and location of the hidden dystrophic painful left hand. He looks inside the mirror and sees the reflection of the normal hand. Suddenly his hand looks normal, no longer swollen. That's obvious because he's looking at the reflection of the normal hand, and it looks like you resurrected his normal hand in the mirror, and it's optically superimposed in the position of the dystrophic swollen hand.

Now you ask him to send signals to both hands as if he were moving them, clenching and unclenching or rotating while he's looking in the mirror. Now he's going to get the impression—you don't initially ask him to actually move the left hand because if he moved it would be painful, he only moves his right hand—and he imagines his left hand moving. What then happens is that the patient gets the visual image that his left hand, which is immobilized and paralyzed, is again obeying the brain's command—it looks like it's moving and is not painful. This way you unlearn the learned pain and the learned paralysis. The astonishing thing is that the hand actually does start moving for the first time in his life, first time in decades, first time in years. It works better if you do it very soon after the dystrophy sets in, a few weeks or months afterward. The hand starts moving again and the pain subsides, and in a remarkable example of mind/body interaction, the swelling also subsides, often in a matter of hours.

This chronic pain disorder is considered intractable, incurable. It has been known for decades. I think it was discovered more than 100 years ago, for which people have done dorsal rhizotomy, cut the nerves going to the spinal cord, cut the spinal cord to treat it. They do a sympathetic ganglionectomy that does work to some limited extent. You can treat it equally effectively, if not more effectively, with just a two-dollar mirror. The patient looks inside and moves his normal hand. We suggested this therapy some years ago, but it was actually Candy McCabe who first properly described it. We suggested the idea, but she discovered it independently.

There have been clinical trials on this from a group in Germany, I believe, on 50 patients. The discovery was originally made on a handful of patients. Since then there have been double-blind, placebo-controlled crossover trials, which is the best type of clinical trial you can do, and people have found dramatic recovery from this pain in a matter of a few weeks of mirror treatment. Then the pain stays gone for a period of at least six months, and then you may need a refill after that. Imagine the amount of pain and agony and invasive surgery this has saved. Sometimes you come up with an off-the-wall half-plausible hypothesis and there can actually be a clinical use for it. One example is RSD, or reflex sympathetic dystrophy (now called Complex Regional Pain Syndrome).

We've talked about synesthesia, we've talked about apotemnophilia, and we've talked about reflex sympathetic dystrophy. There are other syndromes like this that we've studied. Another syndrome is called Capgras syndrome, where a patient has been in a coma for a week or two and comes out of the coma. He's a little bit slowed down. He has mild dysarthria and problems talking, but is otherwise mentally perfectly lucid and normal and can hold a fluent conversation, can play chess with you, can do arithmetic. Everything seems fine except a little bit of slurring of speech. This chap looks at people and can recognize them, no problem. He's not psychotic, not mentally disturbed. The conversation is normal except for the little bit of slurring.

He looks at his mother and says, "Doctor, this woman, you know, she looks just like Mother, but she's not. She's a stranger, some other woman who looks like my mother, but in fact is not my mother. She's an imposter." Sometimes it develops a paranoid touch. He says, "Why is this woman following me all the time? She's not my mother. She's pretending to be my mother."

Why does this happen? The Freudian explanation again—by the way, I don't mean to do too much Freud bashing. I know it's fashionable in New York, but I think that he had deep insight into the human condition, especially the role of the unconscious, which we are increasingly realizing is largely true, and Eric Kandel has written about this. But anyway, it's fun to do—so some Freudians had a theory about Capgras syndrome that when this chap was a young baby, when he was an infant, he had a strong sexual attraction to his mother. Freud called it the Oedipus Complex. As he grew up, the cortex developed and started inhibiting these latent sexual urges toward his mother and therefore, as an adult, he's no longer sexually turned on by his mother. But then a blow to the skull damages the cortex and these flaming sexual urges come to the surface of consciousness and suddenly and inexplicably he finds himself sexually aroused by his mother. He says, "My god, this is my mom. How can I be sexually turned on? This must be some other strange woman."

This is, of course, a very ingenious idea, as indeed a lot of Freudian ideas are. It doesn't work because I've seen patients, at least one patient, who had the same delusion about his pet dog, pet poodle, saying, "Doctor, this is not Fifi. It's some other dog pretending to be Fifi." Now if you try to apply the Freudian analysis to this you've got to start talking about the latent bestiality in all humans and some rubbish like that. So it doesn't work. This got me thinking that there's something going on that's probably neurological in the brain.

I'm mainly an experimental scientist and we go with the flow. It's like charting the source of the Nile. You don't know when the next surprising twist and turn is going to be. It's a great adventure. A grand love affair with nature with all these twists and turns and unpredictable events. That's how we do experiments, but all of it is headed toward the goal of understanding human nature, but understanding it piecemeal. For example, you can't ask, "What is consciousness?" Some people do, but it's too nebulous an idea. In fact, philosophers have criticized this approach. But I think it's okay to ask questions like Francis Crick did.

Well, what is consciousness? Philosophers like Colin McGinn and others have argued that this is utterly mysterious. The human brain can never comprehend itself, and can certainly not comprehend mysterious phenomena like consciousness. Somebody like Crick would vehemently disagree. And I would agree with Crick.

Crick and Koch, for example, have argued that there is a structure called the claustrum that is a thin layer of tissue underlying the insular cortex of the brain. What's exciting about this layer of tissue, what caught Crick's eye and Koch's eye, was that it doesn't have any known function, unlike other regions of the brain. There are many regions that we don't know the function of, but the claustrum is especially mysterious. It's not a tiny, little structure. It's a medium-sized structure, and it's homogenous in its cell constituents. It also doesn't have the layered structure as with the rest of the cortex.

The astonishing thing Crick noticed was that it's connected to almost every part of the brain, including every part of the cortex. It seems reciprocal. It sends connections to the somatic sensory cortex and receives connections back from the somatic sensory cortex. It sends signals to the amygdala, back from the amygdala, to the anterior cigulate, back from the anterior cigulate. In fact, it's very hard to find any region of the brain that is not connected to the claustrum. John Smythies, in our lab, and I have now picked up the gauntlet where he left it.

Crick, for example, has been rewarded in the past for analogies, for big metaphorical leaps. I don't think he actually says this, but if you look at the double helix and the complementarity of the helix, the two sides of the helix, we're struck by the analogy between this and the complementarity between parent and offspring. There's a huge leap of faith there. He says, why do dogs give birth to dogs and not to pigs? Any child will ask this question; you and I won't. But Crick asks that question—why do dogs give birth to dogs and not to pigs? There's a complementarity between offspring and parent. Might it be the case that the complementarity of the two strands of the helix actually dictates complementarity of offspring and parents? This was the big leap. Then, of course, he figured out the genetic code and modern biology was born. He's primed to think in terms of linking seemingly unrelated phenomena, of linking structure and function.

Then he approaches the claustrum and he says, What's the most fundamental thing about consciousness? So axiomatic, in fact, that you take it for granted? That is the fact that you are one person—unity of many attributes of human consciousness. The continuity. The time travel—the ability to go to and from in time—looking into the future, visiting nostalgic memories from the past, stringing them together in approximately the right sequence. Laughter is uniquely human, and we can't imagine laughing without being conscious. Many attributes of human conscious experience—self-awareness is another attribute. Putting it crudely, consciousness is aware of itself.

Now, the central attribute of human conscious experience, so fundamental, in fact, that we take it for granted, don't pause to think about it, is the sense of unity. You've got a diversity of sensory experiences. You see things, you listen to things. This harks back to what I was saying about synesthesia. You taste things. You have hundreds of memories throughout a lifetime. Yet you think of yourself as a unified person. Yet all of these happen to you. You, John, or me, Rama. It all happened to me and I'm one person. Despite this diversity of sensory experiences, this bewildering sensory cognitive blitz of memories and sensory impressions, I experience unity. How does that come about?

Another way to formulate this question is that there are different brain regions actively processing different aspects of information, including memories, and yet you experience yourself as a unity. Many philosophers will argue that this is a pseudo problem, not a true problem. In fact, Crick adopts the opposite view; he and Koch debunk the idea that it's a pseudo problem. He says that the most axiomatic thing about consciousness is its unity. And guess what the claustrum is doing? It's getting sensory inputs, even inputs from the motor cortex. It's getting inputs from every region of the brain in one little gathering place and sending messages back. It's ideally suited for performing this unifying role.

There's an analogy here between what the structure of the claustrum is and what the phenomenology of consciousness is. Maybe this is not just a superficial analogy. Maybe it's deep. Maybe the clue to consciousness lies in looking at the structure of the claustrum, a detailed study of its microanatomy and its connections to the rest of the brain.

Questions of that nature, trying to explain functions like consciousness, like self-awareness, like qualia, in terms of brain structures, is something that Crick pursued, and I think it's something that I'd like to pursue as well, and we have been trying. We all share his agenda—though obviously not his stature.

6

The Social Psychological Narrative—Or—What Is Social Psychology, Anyway?

Timothy D. Wilson

Sherrell J. Aston Professor of Psychology, University of Virginia; author, *Strangers to Ourselves* and *Redirect*.

INTRODUCTION by Daniel Gilbert

Psychology has always had a love-hate relationship with the unconscious, but mainly hate. The unconscious was the cornerstone of Freud's theories about the mind, but William James expressed the views of many early-20th-century scientists when he referred to it as "the sovereign means for believing what one likes in psychology, and for turning what might become a science into a tumbling-ground for whimsies." James's antipathy was contagious and his arguments won the day. The unconscious was banished to psychology's basement for more than half a century.

But in the mid 1970s, Tim Wilson and Dick Nisbett opened the basement door with their landmark paper entitled "Telling More Than We Can Know," in which they reported a series of experiments showing that people are often unaware of the true causes of their own actions, and that when they are asked to explain those actions, they simply make stuff up. People don't realize they are making stuff up, of course; they truly believe the stories they are telling about why they did what they did. But as the experiments showed, people are telling more than they can know. The basement door was opened by experimental evidence, and the unconscious took up permanent residence in the living room. Today, psychological sci-

ence is rife with research showing the extraordinary power of unconscious mental processes.

If liberating the unconscious had been Wilson's only contribution to psychological science, it would have been enough. But it was just the start. Wilson has since discovered and documented a variety of fascinating ways in which all of us are "strangers to ourselves" (which also happens to be the title of his last book—a book that Malcolm Gladwell, writing in The New Yorker, correctly called the best popular psychology book published in the last 20 years). He has done brilliant research on topics ranging from "reasons analysis" (it turns out that when people are asked to generate reasons for their decisions, they typically make bad ones) to "affective forecasting" (it turns out that people can't predict how future events will make them feel), but at the center of all his work lies a single enigmatic insight: we seem to know less about the worlds inside our heads than about the world our heads are inside.

The Torah asks this question: "Is not a flower a mystery no flower can explain?" Some scholars have said yes, some scholars have said no. Wilson has said, "Let's go find out." He has always worn two professional hats—the hat of the psychologist and the hat of the methodologist. He has written extensively about the importance of using experimental methods to solve real-world problems, and in his work on the science of psychological change—he uses a scientific flashlight to chase away a whole host of shadows by examining the many ways in which human beings try to change themselves, from self-help to psychotherapy, and asks whether these things really work, and if so, why? His answers will surprise many people and piss off the rest. I predict that this new work will be the center of a very interesting storm.

Questions that I have asked myself throughout my career are largely ones about self-knowledge and the role of the conscious mind versus unconsciousness; the limits of introspection; and the problems of introspection—for example, how it can sometimes get us into trouble to think too much about why we're doing what we're doing. These are questions I began asking in graduate school with my graduate adviser, Dick Nisbett, and they have concerned me ever since.

There has been a question lurking in the back of my mind for all those years, which is, how can we take this basic knowledge and use it to solve problems of today? I grew up in the turbulent 1960s, in an era when it seemed like the whole world was changing, and that we could have a hand in changing it. Part of my reason for studying psychology in the first place was because I felt that this was something that could help solve social problems. In graduate school and beyond I fell in love with basic research, which is still my first love. It is thrilling to investigate basic questions of self-knowledge and consciousness and unconsciousness. But as those other, more applied questions have continued to rattle around and recently come to the fore, the more I realized how much social psychology has to offer.

One of the basic assumptions of the field is that it's not the objective environment that influences people, but their constructs of the world. You have to get inside people's heads and see the world the way they do. You have to look at the kinds of narratives and stories people tell themselves as to why they're doing what they're doing. What can get people into trouble sometimes in their personal lives, or for more societal problems, is that these stories go wrong. People end up with narratives that are dysfunctional in some way.

We know from cognitive-behavioral therapy and clinical psychology that one way to change people's narratives is through fairly intensive psychotherapy. But social psychologists have suggested that for less severe problems, there are ways to redirect narratives more easily that can have amazingly powerful long-term effects. This is an approach that I've come to call story editing. By giving people little prompts, suggestions about the ways they might reframe a situation or think of it in a slightly different way, we can send them down a narrative path that is much healthier than the one they were on previously.

One of the first studies I did after graduate school tested a story-editing intervention of this kind. We recruited a sample of college students who were caught in a self-defeating thought cycle, where they were not doing well academically (these were first-year students) and were quite worried. They seemed to be blaming themselves and thinking that maybe they were one of those admissions errors that just couldn't cut it at college, which of course made it all the more difficult to study.

We did a brief intervention where, in about 30 minutes, we gave them some facts and some testimonials from other students that suggested that their problems might have a different cause: namely that it's hard to learn the ropes in college at first, but that people do better as the college years go on, when they learn to adjust and to study differently than they did in high school and so on.

This little message that maybe it's not me, it's the situation I'm in, and that that can change, seemed to alter people's stories in ways that had dramatic effects down the road. Namely, people who got this message, as compared to a control group that did not, got better grades over the next couple of years and were less likely to drop out of college. Since then, there have been many other demonstrations of this sort that show that little ways of getting people to redirect their narrative from one path down another is a powerful tool to help people live better lives.

Another issue that interests me is that a lot of the existing interventions out there to help people are not based on theory and, even worse, haven't been tested. If there's one thing social psychologists do know how to do, it's how to do experiments and how to test whether an intervention is working, and with good control groups and statistical analyses, seeing whether something works or not. Yet a lot of the current programs in a wide variety of areas have never been vetted in that way, and are just based on common sense.

There are lots of famous examples—for example, the D.A.R.E. antidrug program, which my two kids went through when they were in school. In fact, 70 percent of schools in America use this program. It was not tested until fairly recently, and the results showed that not only does it not work, but there is also a hint of evidence that it actually increases alcohol and tobacco use in students. I find it shocking that something that turns out to have a negative effect, or at best, no effect, has been implemented in 70 percent of our schools before we even tested it.

There are lots of other examples. Scared Straight programs to scare at-risk kids out of a life of crime turn out to increase the likelihood they will commit crimes. Yet Scared Straight programs are still in use in many communities in the United States. There is a program intended to prevent child abuse, called Healthy Families America, that has been implemented throughout the United States at a cost of millions of dollars. It turns out to have no effect.

Then there's the whole self-help industry, which is interesting because it's not that all of their messages are wrong; it's that they're packaged in a way that give people hope that isn't backed up by science. I tend to think of the self-help industry as kind of like playing the lottery. That is, if we buy a lottery ticket, we're buying hope. We don't really think we're going to win, but for the week before the drawing, we can dream that we're going to suddenly have millions of dollars.

Self-help books are a little bit like that, where we buy them with a promise that our lives will suddenly be better, and all our problems will be solved. We kind of know that's probably not true. But we have a little bit of hope that it will come about.

In fact, there's something in that industry called the 18-Month Rule, that the person most likely to buy a self-help book is one who's bought one 18 months earlier. All of this is a little galling to a social psychologist: there actually is some pretty good research on how to become happier and how to overcome personal difficulties that can be done relatively simply but which the self-help industry ignores. For example, my friend and colleague Jamie Pennebaker has developed a writing exercise that is typically done three or four nights in a row, where you write about a problem for about 15 minutes each time. Doing so has remarkable long-term benefits on people's health and well-being.

Researchers such as Ethan Kross and Ozlem Ayduk have honed this method and, along with Pennebaker, have shown how it works. Think back to the story-editing metaphor: what these writing exercises do is make us address problems that we haven't been able to make sense of and put us through a sense-making process of reworking it in such a way that we gain a new perspective and find some meaning, so that we basically come up with a better story that allows us to put that problem behind us. This is a great example of a story-editing technique that can be quite powerful.

Let's go back to a basic question that often comes up: namely, what is social psychology? It's a good question, because it's something the public doesn't really understand. For example, there was a recent article in the *New York Times* by an economist who referred to me and my friend Dan Gilbert as economists. Those are fighting words! But in a way it is our field's own fault that we haven't succeeded in making social psychology more a part of the public discourse.

Social psychology is a branch of psychology that began in the 1950s, mostly by immigrants from Germany who were escaping the Nazi regime—Kurt Lewin being the most influential one. What they had to offer at that time was largely an alternative to behaviorism. Instead of looking at behavior as solely the product of our objective reinforcement environment, Lewin and others said you have to get inside people's heads and look at the world as they perceive it. These psychologists were very influenced by Gestalt psychologists, who were saying the same thing about perception, and they applied this lesson to the way the mind works in general.

The other big contribution of these early social psychologists was a methodological one. There were experiments at the time, but largely on things such as perception and memory. The idea that you could also do experiments on more complex issues about social interaction and social influence was quite novel at the time. Lewin and his students and colleagues, such as Leon Festinger, Hal Kelley, Stanley Schachter, and others, showed that you use rigorous scientific methods to study how the mind works more broadly in a social context.

But to be honest, the field is a little hard to define. What is social psychology? Well, the social part is about interactions with other people, and topics such as conformity are active areas of research. Everyone knows the famous Milgram studies on obedience to authority, which showed that a strong authority figure can lead others to shock people to the point of death, in their minds.

There's also a branch called cognitive social psychology, which overlaps with cognitive psychology and research on judgment and decision making, focusing on mental processes that occur in a social context. How does the mind work and how does it make decisions? How do people think about themselves and the social world? Social psychologists have a unique way of looking at the mind, doing so very broadly and considering the role of emotion, instead of focusing solely on cold cognition. People like me, Dan Gilbert, and many others are investigating social cognition as a way of understanding how people think about themselves and the social world and how this influences their behavior. For example, Dan and I have been looking at the topic of affective forecasting, which is concerned with the way in which people think about the future and how they think they will react emotionally to a specific event that might befall them.

How will we feel and how long will we feel that way if we become ill, or if we have a windfall of money, or if we take this career path, or if we marry this person instead of that person? Many of our most important decisions in life are based on these affective forecasts, whereby we try to gauge how we will feel about an event in the future, especially over the long run. We are by no means terrible at this; obviously we have a pretty good sense of what will lead to positive feelings and what will lead to negative feelings. But there are systematic mistakes to which people are prone when making affective forecasts.

Perhaps the most common is what we call the impact bias, which is that people overestimate the emotional impact of many events on their lives. We think that if we win the lottery we'll be happy forever. The research on that suggests that not only is that not true, but if anything, lottery winners become less happy, often, because their lives are disrupted in any number of ways. On the negative side, we tend to think that those things that we dread, that would be awful, the death of loved ones, the loss of a job, and so on, will make us unhappy forever. Although they are terrible things to endure, we are more resilient than we anticipate

Timothy D. Wilson

and often get over these events more quickly than we anticipate.

This research has been picked up by medical researchers looking at treatment decisions that often are very difficult choices to make, such as how to treat a disease when there are various options available. People often choose treatments that they think will lead to the best overall well-being and quality of life, but they are not necessarily correct. Physicians struggle with how to educate people that, in their experience, option A may be better than B, but people are convinced that B is better than A, in terms of quality of life.

Legal scholars have begun to look at affective forecasting in terms of decision making by juries and others, when the goal is to try to gauge the impact of events on a plaintiff's overall well-being, as is often done in civil suits.

But research on affective forecasting has some life lessons for all of us. I have to confess that I come from a family of worriers; my family was one of those that always imagined the worst scenario that could happen, and ruminated on the fact that gee, you know, that's going to be terrible. Research on affective forecasting has been a solace because I know that yes, terrible things might happen, and if they do, it will be terrible at first, but then life goes on. We are pretty resilient creatures, and sooner rather than later, we'll find a way to deal with life's worst blows.

Social psychologists think they have something to offer in terms of the discourse on how to solve problems, how to think about many of the issues facing us today. But we haven't done a very good job of making our field part of that discourse with policy makers. Many policy makers, if they're thinking about a problem, such as how to increase condom use in Africa, how to reduce poverty in the United States, or how to reduce prejudice and stereotyping, turn to economists. Some of my best friends are economists; I don't mean to disparage the field, but they think differently about these problems than do social psychologists. They think of human behavior as largely governed by external incentives. Most economists don't take the social psychological approach of trying to get inside the heads of people and understanding how they interpret the world.

I'll give one example. When economists think about how to solve a problem such as closing the achievement gap in education, or reducing teenage pregnancy, their inclination is to use incentives. What if we pay people to do well in school, give kids money to study and to get good grades? Or what if we take girls who are at risk for becoming pregnant and pay them a dollar for each day they are not pregnant?

To a social psychologist, it is a little naïve to think that adding external incentives is all you have to do. Not to say that incentives can't work, but they can sometimes backfire if you look at it through the eyes of the person who is getting that incentive. There's some research in social psychology suggesting that external incentives can undermine intrinsic interest in an activity because people begin to think that the only reason they're doing it is for the money. That erodes any interest in that activity there was to start with.

Now, in defense of economists, they do think big. They think of systems at large, and they are increasingly doing experimental interventions to try to see what works. I'm envious, in many ways, of the scale at which they are able to attempt to solve problems. One limitation of social psychology is that we have concentrated mostly on basic issues about the mind in laboratory studies, many of them done with college students. There's an issue in how to scale up those interventions to see if they work with society at large. That's a gap we need to fill, and people are increasingly doing so. Taking some of these basic social psycho-

Timothy D. Wilson

logical principles and seeing if they can be scaled up to work more generally, not just in the laboratory with a college student, is an exciting prospect, and there have been some spectacular successes in doing just that.

I'll give one example that I'm very fond of that was done by Geoffrey Cohen and his colleagues. Geoff is a social psychologist who trained with Claude Steele, and was very well versed in Steele's self-affirmation theory, which is the idea that when we feel a threat to our self-esteem that's difficult to deal with, sometimes the best thing we can do is to affirm ourselves in some completely different domain. If I am concerned that I can't make it in academics, it can take the heat off that concern if I think of something I'm very good at and I care about in some other domain, such as that I'm a family man or interested in politics or what-have-you. Research on self-affirmation theory was done mostly with college students in the laboratory, showing that affirming oneself in an unrelated domain is a powerful way to restore self-esteem.

Well, Cohen had the insight that maybe this can be used in middle schools with minority kids, African-American kids in this case, who are facing challenges to their self-esteem in the academic realm. The negative stereotype that African-Americans are not as smart as whites has very damaging effects on African-American students, because when they are in academic achievement situations, not only do they have to worry about doing well for their own sake, they have this extra baggage of, gee, if I don't do well, I'm also confirming a negative stereotype about my group. This is what Claude Steele has referred to as stereotype threat.

Well, Cohen thought, maybe we can take the research on self-affirmation and use it to reduce stereotype threat in middle school kids. If we can get African-American middle school kids to affirm themselves in some domain unrelated to the academic realm, he hypothesized, this will take the heat off and make it actually easier for them to do well academically. He did an intervention in which middle school kids wrote about a value that they cared about in their lives, other than academics. They did this for 15 minutes, three to five times during the semester, depending on the version of the study. That was it: write about something you care about in your life other than academics.

This was a good experiment, because there was a randomly assigned control group of kids who did not do this exercise. The intervention had remarkably long-term effects: the African-American kids who did the writing exercise, compared to the control group, did better academically for the next couple of years. In fact, the intervention closed the achievement gap between the black and white students by 40 percent. It had no effect on the white kids because they weren't at risk for stereotype threat in academic domains. It seems to have lowered the heat for the black kids when they were in an achievement situation, enabling them to do better.

That's one of many examples in which a social psychological principle discovered in the laboratory with basic research can be scaled up to solve a problem more generally in education or other domains. My dream is that policy makers will become more familiar with this approach and be as likely to call upon a social psychologist as an economist to address social issues. Now, again, my field deserves some of the blame for this; historically we haven't always met policy makers halfway and tried to do more of this kind of work. But that's changing, and social psychological interventions are becoming more and more common.

In addition to using social psychological research to inform policy, it can be used to change the way we live our own lives. Something I think about a lot is how my life has changed as a student in this field. The importance of how we frame things and the idea that we have to look at our construals of the world—well, it's a little unsettling, personally, because it does suggest that the way we view the world isn't necessarily the correct way or the best way. Or that what we're seeing isn't necessarily reality. And that can lead to a little less confidence or certainty about the world and our place in it. But it's not such a bad thing to have a little humility that our view may not be the only one, or even the best one.

Another interesting question is the role of evolutionary theory in psychology, and social psychology in particular. I can trace the history of that a little bit. I got my PhD in 1977, and that was just around the time that evolutionary theory was being applied to social behavior, through E. O. Wilson's work in sociobiology, and others'. But it really, at least right away, did not make inroads into social psychology. In fact, in the late '70s, if a social psychologist were to say I'm an evolutionary theorist, that would have been a really taboo thing to say. It would have struck people as overly deterministic, and perhaps even sexist, to look at gender differences in social behavior as somehow inherent in the human condition.

But things sure have changed. Evolutionary psychology has become a dominant force in the field. There are many who use it as their primary theoretical perspective, as a way to understand why we do what we do. I'll get myself in trouble with some of my colleagues for saying this, but I am not a fan.

Evolutionary theory has its use. Of course, evolution is true, as a general theory of how the human species evolved. As an explanation for current social behavior, it can be a useful heuristic, if it can generate hypotheses that we would not have come up with otherwise that can then be tested with rigorous methods. But too often, there's a very loose kind of theorization that goes on, where people just tell a story and assume that it's true because it kind of makes sense.

I've been writing this paper in my head for many years, that maybe I'll put on paper at some point, called "Evolutionary Theory: The New Psychoanalysis." There are some striking parallels between psychoanalytic theory and evolutionary theory. Both theories, at some general level, are true. Evolutionary theory, of course, shows how the forces of natural selection operated on human beings. Psychoanalytic theory argues that our childhood experiences mold us in certain ways and give us outlooks on the world. Our early relationships with our parents lead to unconscious structures that can be very powerful.

But both theories led to a lot of absurd conclusions, and both are very hard to test rigorously. The influence of psychoanalysis waned in research psychology because it was too broad. It made too many assumptions that were very hard to test, and basically it explained everything. That said, it did actually lead to some interesting hypotheses that were tested rigorously. One example is Susan Andersen's work on transference, which shows that, indeed, we do have blueprints about relationships that form and influence our perceptions of new relationships.

Evolutionary theory, in a way, has the same status. It can explain virtually anything. It can be a useful heuristic, as I mentioned. But at the same time, I think it is way too broad. Another parallel between the two theories is that they both seem obsessed with gender differences. There are many ways that we could think about human behavior, but zeroing in on why women are different from men is something both theories were obsessed with, and both theories have gotten wrong, to some extent, in attributing differences in social behavior to genetic hardwired influences.

The storytelling method is a real problem. In one of Steve Pinker's books he gave the example that not everything is an ad-

Timothy D. Wilson

aptation. For example, the fact that blood is red, he says, isn't necessarily a result of natural selection. Well, I could make up a story as to why it is. What if in our very early mammalian history, blood was more brown, but there was a mutation that made it more red, and that turned out to have survival value because if an animal were bleeding, those with red blood would be more likely to notice it, and then they'd lick it. Because licking has healing properties, this conveyed a survival advantage, and so red blood was selected for, and blood became red. Am I right? Or is Steve right, that the color of blood is not an adaptation? Who knows. The plausibility of a story is not a good way to settle a question scientifically. Now again, in fairness, there are some very interesting hypotheses that we would not have come up with if it were not for evolutionary principles that have led to some interesting lines of research. But there are not that many of them.

One example where evolutionary psychology led to some interesting testable hypotheses is work by Jon Haidt, my colleague at the University of Virginia. He has developed a theory of moral foundations that says that all human beings endorse the same list of moral values, but that people of different political stripes believe that some of these values are more important than others. In other words, liberals may have somewhat different moral foundations than conservatives. Jon has persuasively argued that one reason that political discourse has become so heated and divisive in our country is that there is a lack of understanding in one camp of the moral foundations that the other camp is using to interpret and evaluate the world. If we can increase that understanding, we might lower the heat and improve the dialogue between people on opposite ends of the political spectrum.

Another way in which evolutionary theory has been used is to address questions about the origins of religion. This is not a literature I have followed that closely, to be honest, but there's obviously a very interesting discourse going on about group selection and the origins and purpose of religion. The only thing I'll add is back to what I've said before about the importance of having narratives and stories to give people a sense of meaning and purpose—well, religion is obviously one very important source of such narratives. Religion gives us a sense that there is a purpose and a meaning to life, the sense that we are important in the universe, and that our lives aren't meaningless specks like a piece of sand on a beach. That can be very powerful for our well-being. I don't think religion is the only way to accomplish that; there are many belief systems that can give us a sense of meaning and purpose other than religion. But religion can fill that void.

Timothy D. Wilson

7

The Adolescent Brain

Sarah-Jayne Blakemore

Royal Society University Research Fellow and Professor of Cognitive Neuroscience, Institute of Cognitive Neuroscience, University College London, UK.

INTRODUCTION by Simon Baron-Cohen
Psychologist; Professor of Developmental Psychopathology and Director of the Autism Research Centre, Cambridge University; Fellow of Trinity College, Cambridge; author, *The Essential Difference* and *The Science of Evil*.

Sarah-Jayne Blakemore is a leading social neuroscientist of adolescent development. She has reawakened research interest into the puberty period by focusing on social cognition and its neural underpinnings. Part of her question is whether adolescence involves egocentrism, as many popular conceptions suggest, since this is testable.

Part of her originality is to remind us of the remarkable changes in brain structure during adolescence, given that the traditional focus of developmental psychology is on early childhood. Using a range of techniques, including conducting elegant MRI studies, she illuminates a neglected phase of cognitive development. Given that the sex steroid hormones are produced in higher quantities during this period, her research opens up interesting questions about whether the changes in the brain are driven by the endocrine system, or by changing social experience, or an interaction of these factors.

I'm particularly interested in the development of the adolescent human brain. The reason I became interested in the adolescent brain is twofold. Firstly, we know that most adult mental disorders have their onset at some point during the teenage years, so if you look at disorders like anxiety disorders, depression, addictions, and eating disorders, almost all of them will have their onset sometime during the teenage years.

Schizophrenia, as you might know, is a very horrific psychiatric condition that's characterized by delusions, like being paranoid and thinking that people are out to get you, and hallucinations, like imagining that people are talking to you inside your head, hearing voices. That has its onset at the end of adolescence, normally in the early 20s, on average. So that's one reason I think it's really important to study the adolescent brain. The hypothesis is that something is going wrong in normal brain development to trigger these psychiatric and psychological disorders.

The second reason adolescence is an interesting period of life to study is because unlike most other periods of life, the leading causes of death in adolescence are accidents. That's the number one leading cause of death during the period of adolescence; the second is suicide. The accidents are caused, generally, by risk taking. So we know that teenagers take more risks than either children or adults. The question is, why? Why is adolescence associated with this phenomenon like increased risk taking, and especially when adolescents are with their peers? So peers become really influential in adolescence. Adolescents are driven toward impressing their peers, trying to seek approval of their peers, and becoming more and more independent from their parents. Social cognition, the social brain, seems to change during the period of adolescence, and that's something that particularly interests me. And finally, self-awareness: awareness of one's self, and consciousness of one's

Sarah-Jayne Blakemore

self. We all know, if you remember what it's like being a teenager, that feeling of heightened self-consciousness that seems to happen in early adolescence where you become easily embarrassed by things like your parents, or social situations where you're not seen as cool, and that kind of thing.

That's what we're interested in looking at, the development of these kinds of cognitive processes like self-awareness, social understanding, the understanding of other people, and risk taking and decision making during this period of life.

In our research, what we're really interested in is tracking the development of the brain, both in terms of its structure and its function, and also behavior during the period of adolescence, and comparing that in typically developing adolescents to healthy adults, and we are also moving toward looking at adolescents who develop psychiatric or psychological disorders—in particular, adolescents who go on to develop schizophrenia—so one of my overarching interests in my ultimate long-term goals is to look at the brain development patterns in teenagers who later develop schizophrenia.

The kinds of experiments we do involve brain imaging. We use structural MRI and functional MRI, where we bring teenagers into the lab and scan their brains and acquire structural images of their brains, and also look at how their brains function, so we record where activity is being produced in their brains during certain tasks. We might give them a task which involves thinking about other people, or thinking about themselves, or taking risks or making decisions, and compare brain activity in the average adolescent brain during those tasks with the average adult brain.

This is a kind of question that many labs around the world are now starting to ask. But the interesting thing is that this is a very, very recent field, and if you look in the literature, 15 years ago you would have found virtually nothing. Nothing was known about how the living human adolescent brain develops. We've now learned a great deal in the last ten years, due to advances in imaging technology, so we now have the ability to scan the living human brain and track its development, both in terms of structure and function across the life span, and that has taught us really a great deal. It's really revolutionized what we understand, what we know about how the living human brain develops.

One of the main findings from this kind of research is from a very large pediatric neuroimaging project at the National Institute of Mental Health in Bethesda, in the United States, where they have scanned children, adolescents, and adults, many thousands of individuals. As they get older they bring them back into the lab for a scan every couple of years, and they now have approximately over 8,000 scans from 2,000 individuals, and they put all this data together to form this kind of semi-longitudinal data set, which has shed a great deal of light on how the brain develops structurally during this period of life.

One of the things they found—and they found many different findings from this data set—is that the human cortex undergoes much more protracted development both in terms of gray matter and white matter volumes than was ever previously thought. We know, firstly, from this data set, that in many different cortical regions, gray matter—which is mostly found in the cortex, the surface of the brain, and contains cell bodies and synapses, the connections between cells—gray matter increases during childhood, peaks at some point during mid- to late childhood or early adolescence in most cortical areas, and then declines really dramatically during the period of adolescence right into the 20s or even the 30s. We don't quite know what this corresponds to be-

Sarah-Jayne Blakemore

cause MRI doesn't have the resolution to tell us about what's going on at a cellular level, or a synaptic level, but we know from postmortem human brain tissue studies that there is a large amount of synaptic pruning—that is the little connections between brain cells, they're called synapses—and in development they firstly increase in number and then decrease again, and that decrease in synapses is caused by synaptic pruning, where excess synapses are just eliminated—synapses that aren't being used are pruned away or eliminated. That we know from postmortem human brain tissue studies. It continues during the period of adolescence, right into the 30s.

We think that the decrease in gray matter volume during the period of adolescence, that we've gleaned from these living human brain MRI studies, corresponds to synaptic pruning going on during the period of adolescence. At the same time, there's an increase in white matter across the brain, and this is thought to be due to the fibers of cells, so these are axons along which electrical impulses pass from cell to cell in the brain, become coated in a white substance called myelin during development. We know that again from animal studies, from cellular studies, and this appears in MRI scans as an increase in white matter.

Now, the functional consequence of that is that myelin acts as an insulator and speeds up the transmission of signals from cell to cell, and so we think that the implication of that in terms of its function is to speed up signaling between brain regions, and that continues to happen throughout the first, at least the first three, even four, decades of life.

Those are the structural studies that have revolutionized what we know about the development of the living human brain. Many labs around the world, including mine, look at how the brain functions in adolescence, compared with adults, in a variety of tasks. I'll just give you an example. One of the areas of cognition that we're particularly interested in is social cognition: the social brain, how we understand other people and how we interact with other people. We know from many human functional MRI, or fMRI, studies that the social brain is a network of brain regions that is consistently activated whenever adults think about other people. There are about three different regions in the brain, one in the medial prefrontal cortex and two other regions in the temporal lobe: the posterior-superior temporal sulcus and the anterior temporal cortex. It doesn't really matter about the names, but the point is that that network of brain regions in adults is consistently active whenever you think about other people or think about interacting with other people, or think about their mental states or their emotions.

Adolescents use the same network, the social brain network, to a very similar extent, but what seems to happen is that activity shifts from the anterior region, the medial prefrontal cortex region, to the posterior, the anterior temporal cortex or the superior temple sulcus region, as they go through adolescence. In other words, when they're thinking about other people, adolescents seem to be using this prefrontal cortex, right at the front region, more than adults do, and adults seem to be using the temporal regions more than adolescents do.

The question is, why? Why should adolescents who complete the task in the same time and as accurately as adults, why should they be using the prefrontal region of their brain more, and the temporal region of their brain less? That's something that we're looking at now. One possibility is that they're using different cognitive strategies to do these tasks. They're doing the tasks, and even though they're doing them as well, they're doing them in a different way. That's one of the hypotheses that we're currently looking at.

The final strand of our research looks at how behavior changes

Sarah-Jayne Blakemore

during the period of adolescence. If you look back in history over the last 30 or 40 years, if you look at developmental psychology studies of social cognition, what you mostly read is social cognition tasks given to very young children (normally children below the age of five or six). Some of these tasks might be theory of mind tasks. Theory of mind is defined as the attribution of mental states—like the classic example is understanding that someone else can have a belief that is different from your own, or different from reality, and those kinds of tasks.

One of the classic versions of this false belief task is the Sally Anne task. This is a task designed for young children. Normally you have two dolls, Sally and Anne. Sally hides something in a box and then goes out of the room, and when she's out of the room, Anne takes her toy from the box and puts it in a completely different place, a basket. The question is, when Sally comes back into the room, where will she look for her toy? The answer, of course, is that she'll look for her toy exactly where she left it, which was in the box, because she doesn't know that Anne has hidden it somewhere else. Of course, as an adult you know that instantly, but actually until the age of about four, very intelligent, typically developing children tend to get that wrong. They tend not to be able to understand, in this kind of explicit task, that Sally could possibly have a belief that's different from reality, that's different from their own belief.

Given the fact that we know that social brain regions continue to develop, both in terms of structure and function, during adolescence, we were interested in how social cognitive behavior changes in adolescence. It would be weird to think that these parts of the brain can change so dramatically and significantly, and have no consequence on behavior. We were interested in looking at behavior during the period of adolescence; however, when we first started out on this project there were really no tasks that we could use, because almost all theory of mind tasks reached ceiling performance in early childhood. By age five all children get these tasks right almost 100 percent of the time. So we had to find a task that didn't result in 100 percent accuracy.

We used a task that involved taking someone else's perspective in an ongoing kind of communication context. It's very different from a false belief task, but it's more like the way you might use theory of mind in everyday life, constantly having to work out what the other person that you're talking to is intending, what they intend you to understand, and using their perspective in order to guide your ongoing behavior and actions. When we used a task like this, which we know results in many errors, even in adults, what we found was that the ability to take into account someone else's perspective in order to guide ongoing decisions and behavior continues to develop well into late adolescence, so much, much later than what the previous theory of mind, classic theory of mind tasks had indicated.

That's an example, a relatively early example of the kind of results that we found, and we are currently expanding our research into areas like risk taking and peer influence and decision making. Eventually, what we want to do is look at things like, we've started a project to look at things like the genetic influence on brain development during adolescence, how your genes influence how your brain develops during this period of life, and also to look at adolescents who have early-onset schizophrenia, teenagers who start to develop schizophrenia-like symptoms, like hearing voices inside their heads or feeling excessively paranoid.

I did a degree in experimental psychology at Oxford, and this degree in Oxford is very biological. There's a lot of neuroscience

Sarah-Jayne Blakemore

in it. I quickly became interested in the neuroscience aspect; in particular I became interested in schizophrenia. And so for my PhD I decided to study schizophrenia, and I did a PhD in neuroscience at UCL supervised by Chris Frith, who's very well known for his work on schizophrenia, and Daniel Wolpert, who looks at mathematical models of prediction in the brain. I did a PhD on schizophrenia, particularly looking at how people distinguish between self and other, how they make the distinction between when they are causing an action versus when someone else is causing an action. I went on to do a postdoc in France with Jean Decety, also looking at schizophrenia and the perception of causality and contingency.

My interest was always in schizophrenia, and that's why I became interested in looking at the development of adolescence. It was when I was in France that I started to think about why it is that people with schizophrenia tend to develop it in the early 20s. This is a developmental disorder that has its onset relatively late in life, in the early 20s, normally. I became interested in the idea that maybe something in normal brain development goes awry in people who develop schizophrenia. But at that time, ten years ago, looking at the literature, there was very little known about even how the typically developing human brain develops during the period of adolescence, let alone how the brain develops in people who then develop schizophrenia. So that's when I changed track a bit and started to focus my own research on the development of the typically developing adolescent brain.

Now that's becoming a huge area in itself. Many people are asking questions about this area because, I think, it's just a kind of lucky time in history that we are now able to scan the living human brain at all ages, and many people do that across the ages, across the lifespan. Although if you look at the literature, actually, most developmental studies start at about age six, and I think this is because you have to lie completely still inside the MRI scanner to acquire a high-quality resolution, a high-quality image of the brain, and under age six it's really difficult to get kids to lie still enough in the scanner.

I think a lot of labs are focusing on the teenage years because this seems to be a time that is sort of nonlinear in development. It's not just that teenagers represent a kind of continuation of childhood. There's something special about the period of adolescence, where adolescents are driven towards peers and away from their parents. They're driven to develop a sense of self and self-identity, and especially a sense of who they are, how they're seen by other people, in particular their peers. It's a time where there's probably an increased drive to take risks, so from the evolutionary point of view, to sort of move away from the relative security of your family and your parents, and take risks by discovering things for yourself in the outside world.

So we have the ability to scan the brain. There's masses that we don't yet understand about how the brain develops and, in particular, there's a lot that we don't really understand about what the images are telling us, like, what is white matter? What's gray matter? What does it consist of? Why is it changing? We don't know the answers to this, and we won't until we can look at the developing human brain or, I guess, animal brain to a certain extent, in terms of changes that are going on at the synaptic or the cellular level, and mapping that knowledge onto the knowledge that we have from the imaging studies. That's the structure.

Similarly, when we look at how brain activity changes during development, there's a lot we don't understand about that, so when we see a part of the brain change its activity as you get older, we don't know why that is. We don't know whether it might be because that part of the brain is changing structurally. If you have fewer and fewer synapses, for example (connections between cells), or gray matter, as you get older, between early adolescence and early adulthood, then that might result in less and less activity with age, and that's one theory that has been put forward by a number of different labs. That's certainly a possibility. There are many other possibilities as well. It's possible that at different ages you use different brain circuitry to perform the same task because you're using a different kind of cognitive strategy. You might—for example, when you think about social situations as an adult—you might be doing this automatically by just triggering automatically some kind of social script, whereas maybe in adolescence you're more reliant on your own experiences of these situations. It's more effortful and you have to conjure up examples where some similar social situation has happened to you in order to think about the consequences or something like that. There are many possibilities. There are many possible explanations that people are just starting to look at.

There are also methodological issues with scanning the brain at different ages and comparing different groups. For example, the fMRI signal, the BOLD signal—that's the signal of activity that we get from the fMRI scanner—is essentially an indirect measure of neuronal activity. It's an indirect measure of the activity generated by brain cells. It's effectively a measure of blood flow because we know that when neurons are active they require more energy to flow to them in the blood, and that's what we're measuring when we're saying brain activity—what we're really talking about is where blood is flowing to in the brain. For example, if neurovascular coupling—that is, the vasculature that innovates different brain regions—is developing across adolescence, then that will probably affect the BOLD signal that we get from the fMRI scanner, and we just don't really have any data yet on how vasculature changes in the brain across this period of life in humans. There are many, many questions.

My dad is Colin Blakemore. He's a well-known neuroscientist in the UK, and so growing up, our house was very much full of science. We grew up in Oxford a few minutes away from his lab, and we were constantly going to his lab and visiting it. There's a big park next to that, so we'd go to the park and then go to his lab. But on the other hand, when I was at school—high school, that is—I never really felt like, "Oh, I'm definitely going to be a neuroscientist like my dad." And actually what I was really interested in was psychology and developmental psychology, and I did a week's work experience, when I was about 15, with Uta Frith, who, as many people looking at this will know, is a very famous developmental psychologist in the UK who specializes in autism. And I, at the age of 15, went and did a week's work experience with her, where I hung out in her lab in London and watched children with autism being tested on the Sally Anne task, this classic task of theory of mind, and helped develop some spoonerism tests for kids with dyslexia (these are kind of tests where you switch around the first letter of two words—like Jimi Hendrix would be Himi Jendrix or whatever) that Uta was developing at that time. Anyway, that really made me become very interested in psychology, which is why I applied to do a psychology degree at university.

I didn't really explicitly make the connection between psychology and neuroscience until I started my degree, when it became very obvious that if you want to study psychology what you're really studying is the brain. You're already studying the mental process of the brain. And I found that studying the brain was the part of psychology that I was particularly interested in. So I ended up becoming a neuroscientist like my dad.

My sisters—I've got two sisters, and they're not scientists. One is a video-game designer and the other is a pediatric nurse. It's only me who went into my dad's field.

The work I do absolutely relies on interdisciplinary methodology. We work with a team of cognitive neuroscientists. We work with geneticists and physicists and psychiatrists who see patients, for example, with early-onset schizophrenia; pediatricians who specialize in endocrinology when we want to look at, for example, the effects of puberty hormones on brain development. It's almost by definition an interdisciplinary area to work in, and none of the questions that we currently ask could be asked if we didn't collaborate with people from different disciplines, and we help each other to progress the field.

One interesting thing to think about, when you're thinking about brain imaging, is, why is brain imaging important? What does it teach us that we didn't already know from psychology studies? This is a really important question that a lot of people are asking. Why does it matter that we know that one part of the brain is involved with a process? Why does that matter more than just knowing about this process from a kind of psychological point of view? For example, if you know that one method of teaching works better than another method of teaching, so one method of memory rehearsal worked better than another method, why does knowing that the hippocampus is more involved in one than the other—why is that useful? Does it tell you any more than you already knew from the psychology results or the education result? I think this is a very open question and often, actually, especially when you're talking about the implications of neuroscience for education, actually, often it's the case that is sort of seduced by these brain images, and we see them and they are very tangible and people suddenly think, "Oh, my God, it has a biological basis," and they somehow seem more convincing and attractive than just pure psychology results. But often they don't really tell us anything more.

I think the area of adolescent brain development is one of the areas in cognitive neuroscience where actually brain imaging has completely revolutionized what we know. Prior to all these data on the developing adolescent brain, we had a lot of data from psychology and social psychology, and educational research, on how adolescents develop. But these kinds of behavioral changes that adolescents go through were mostly put down to things like hormones, like changing hormones at puberty, and also social environment, like changing social priorities, maybe changing schools, that kind of thing. We just didn't know until 10 or 15 years ago that the brain undergoes such dramatic development and even reorganization during the period of adolescence starting at puberty and continuing right throughout adolescence.

It's that insight that we've gleaned from being able to scan the living human brain that has really changed the way we think about adolescent development, adolescent education, even the legal treatment of adolescence. I think brain imaging there, it's one area in which it has revolutionized how we understand teenagers.

It's interesting to look at the detractors. Actually, there aren't very many, but that's probably because it's a new field and it will just take time for detractors to build up. If we're just focusing in on imaging studies of the adolescent brain, not imaging generally—which has a lot of detractors, but I won't go there—there are people who ask the question, "So what? So what if the brain develops during adolescence?" That's irrelevant when it comes to thinking about educational programs or rehabilitation for teenagers.

I think we do have to be careful about being seduced by these brain images. We know that people tend to be more satisfied when

Sarah-Jayne Blakemore

you give them an explanation that includes some kind of brain term like prefrontal cortex. We've got to be really careful about that, especially when taking the findings from neuroscience to an educational context. That's something I'm really interested in. I'm interested in the links between neuroscience and education. Whenever I talk to teachers or schools, I often end up talking about where the neuroscience has been taken too far, and there are programs that sell themselves on the basis of, well, so-called neuroscience that often makes no physiological sense.

One type of detractor might argue that we can't just scan children's brains or teenagers' brains and show a difference; it doesn't really tell us very much. The fact is, though, that if the brain changes, that must be, arguably, having some influence on cognition and behavior.

The other side of the coin is that this is a period of life where the brain changes very rapidly and dramatically, and some people are starting to think of adolescence as a second critical period or sensitive period in brain development, and if that's the case, then that has really profound implications for the environmental influences on adolescent brain development, how we treat teenagers, and the kinds of social experiences they have. It also means that in principle, things like rehabilitation and interventions targeted at teenagers, for example who might have had negative experiences in early life, should be not a wasted time.

The idea that the brain is somehow fixed in early childhood, which was an idea that was very strongly believed up until fairly recently, is completely wrong. There's no evidence that the brain is somehow set and can't change after early childhood. In fact, it goes through this very large development throughout adolescence and right into the 20s and 30s, and even after that it's plastic forever; the plasticity is a baseline state, no matter how old you are.

That has implications for things like intervention programs and educational programs for teenagers.

There's a second line of questioning which I think is actually really important—I mentioned it earlier on—which is what it means. So talking about the functional imaging changes, scanning the brain and looking at how activity changes across age—what that means in terms of what's the underlying changes in neuronal signatures. We have to be really careful about what we attribute this to, especially because if you get developmental changes in blood vessels and vasculature at this age, then you will probably see that as a change in bold signal arising from fMRI. So we have to be really careful about making assumptions in terms of what we're seeing with the fMRI scanner, and more research needs to be done to understand this.

What's really needed is a better understanding of the technology. If we could scan the living human brain and be able to see brain structures that are cellular, or synaptic levels, that would hugely benefit the field. I have no idea when that's going to be possible, or even how, but putting funding into the technology and the method is really the way of progressing understanding in this field. On the other hand, at the moment funding is so tight that we are very limited in, for example, the numbers of participants. It's expensive scanning humans. You put them in the scanner for an hour and you spend a few hundred pounds just scanning a single subject. So we're limited in the numbers that we can scan, and yet really what we need is much, much larger-scale studies that are longitudinal, so where you're scanning the same individuals across a number of years as they get older. Most of the studies that I've mentioned are not longitudinal, they're cross-sectional, so you're comparing different teenagers with different adults.

It would be ideal if you could scan a very large number of teen-

Sarah-Jayne Blakemore

agers every couple of years as they go into adulthood. The icing on the cake would be to scan a sufficient number of individuals so that you can track people who, for example, develop schizophrenia, and go back and look at their brain imaging data from when they were a teenager, and look at how it differs, how their brain development differs from teenagers who don't develop schizophrenia. That is being done, but it's very, very hard to do that, because as you can imagine, it would take enormous numbers of participants, and it's such a long-term study that you'd really need to be doing it for 20 years.

This is a very new field and we're still learning; we've still got a great deal to learn. In fact, most questions have yet even to be looked at, let alone answered. Things like how the brain develops in adolescents who develop some kind of psychiatric or psychological disorder is something that we know very little about. People are starting to look at that. That's something that I would like to move into. How genes and the environment influence brain development, like for example how adolescent brain development differs between cultures, is something that no one has yet asked, and yet it's bound to. Culture is a huge environmental influence, and we assumed that the environment influences brain development during adolescence, so that's something else that is in the cards for the future to look at.

8
Essentialism

Bruce Hood

Experimental psychologist; Chair in Developmental Psychology
in Society, Bristol Cognitive Development Centre; author,
SuperSense, The Origins of Object Knowledge, and *The Self Illusion.*

I've reached a crossroads in my research and in the questions I'm
now starting to ask. Part of that was driven by some insight and
realization about the direction I was taking, and part of it was also
driven by changes in economic circumstances. Notably, the reduction in funding in this country has impacted upon my field quite
dramatically (behavioral sciences). The way that that has impacted
is that there's far less money to fund research, so the competition
to get funding has become very acute. Now we have to justify with
a view to application. In the past you could just go off on a flight of
fancy, studying the things that were of intrinsic interest.

But now we have to steer our grant applications toward potential application, and certainly we have to write a substantial proportion of the proposal to deal with impact, public engagement.
And that's across the board. As I said, if this had been five, ten
years ago, there would have been some resistance to that, but increasingly now the research councils feel that we, as a public body
funded by taxation, need to be called to account in terms of what
we're doing with the money, the taxpayers' money.

This has led me to start thinking more about what I do in terms
of its tangible application in the real world. That's the external

influences that have been shaping the sorts of questions I'm starting to ask now.

There's been a growing awareness that there have been a lot of problems with the way that psychological research has been going on in the past, very much lab-based type of work. There has been a general issue in the experimental method: what you typically do is you hone in on a question, and you try to refine that question by removing all the extraneous variables to try to make it as clean as possible. But then that does raise the question, to what extent? And does what you eventually find actually have real relevance or validity to the external world? Because in many senses, the complexity of the external world might be part of the problem that the brain is trying to solve.

A number of us have been getting increasingly concerned that the theories that have been derived by purely experimental methods may not necessarily translate into the real world as well. There's a combination of this external pressure to come up with research that is seen to have application, but also a growing concern that maybe some of the findings wouldn't necessarily translate. That's the kind of major forces that have been changing the way that I do things.

I've also got increasingly interested in public communication and education, engagement, transfer of knowledge, and so there's more of that on my plate these days. I did the Christmas lectures in 2011. That was a great opportunity. I've actually changed my job title, I'm professor of developmental psychology in society. I see myself continuing with my research career, but trying to find opportunities to take that out to the general public.

To give you a tangible example, one of the things that we study in my laboratory is how children navigate or build up spatial maps of the world. They wander around our laboratory and try to find targets, which are embedded in the floor as LED switches. It's a very useful experimental measure of foraging. How do you keep track of where you've been? How do you optimize your search behavior? That's led to some interesting ideas and we could have stayed just pursuing that, but now we're starting to say, well, can we see whether those findings actually are relevant to the real world? We're taking those paradigms and testing them out in the local science museum. That provides not only an opportunity to see whether it's relevant in the real world, but also an opportunity to engage the public. It's killing two birds with one stone. It's something that the public can actually do and it seems of importance, contribution, and also trying something that is a little bit more real-world. I'm not suggesting that this is a good absolute model for foraging in the forest or the savannah, but it is a way of taking the work that has, up to this point in time, been done on laptops and then trying to make it closer to what a real-world situation would be.

The environments we have been looking at are at the science museums, and these are dotted around the country, where they're open to the general public, and it's usually a motivated audience that comes in to try to find out what's going on in science, and there's demonstrations. We've seen this as an opportunity to look at some of the findings that we have in environments that are not within the laboratory. So it's not entirely what you would regard as a field study as such, because these are very unusual environments. But it is closer to what would be, certainly a lot farther away from the laboratory settings where most work on intentional search has been done. There are other researchers who actually do experiments in supermarkets and different environments. This is the field of ergonomics, where they're basically trying to find the best way of solving tasks in a work environment, occupational type of way. We're not really attempting to do that. I'm just finding this, we're just looking for more opportunities to engage the public in a kind of activity which has a credible experimental basis.

This is a new venture for us. This is what we're now starting to turn toward. Up to now I've been doing basically lab-based work. This is the new point in the transition I'm going to make, which is to increasingly look for opportunities to do things with the general public. As yet, it's early days. It's driven, as I say, partly by motivation to be seen to be relevant, partly for the fact it's actually cheaper to do research in the real world in terms of getting the numbers through. These are all kind of factors.

The lab has changed, as I said, because the financial situation has changed in the past five years and our field has been hit heavily, I feel, in many ways. Behavioral science and psychology used to be funded by at least seven of the research councils. They've all retracted and there's really only one research council now that funds psychological research. There's greater competition. So that means the labs are getting smaller. We're not taking on as many graduate students, and I think that's probably the right thing because the climate at the moment isn't right. I think it's probably wrong to train our people to a high standard for situations where there aren't actually jobs for them. So that's one issue.

We are doing much more consultancy work. We're doing much more applied applications of the principles that we've been using, the methodologies that we use to study young children, we've been using these in sort of commercial situations. For example, we're just beginning work with Aardman, the animation company. They want questions answered about what children, preschool children and young children, like and find most memorable about some of their projects. Those are empirical questions that you can use our methodologies to unravel. That's an example of an application. The work in the museums, as I say, is the specific example of looking at foraging.

How it's changed the lab? The training that I'm now giving to the graduate students and the postdocs increasingly involves them doing public engagement activity, and that's actually what everyone's asking for. The grant funding bodies, the universities now—as we've seen a change in the funding situation, we're now in this country charging fees, quite substantial fees, and so we're entering into a very competitive situation because all the universities are trying to compete with each other to get the best students. We now have to market ourselves, so there are more of these academics doing videos and snippets and stuff on the Web.

What would I do with a million-dollar chair? Well, my big thing is essentialism. Its origins probably can be traced to the notion of ideal forms, which is a platonic idea. I discovered essentialism basically by reading Susan Gelman's work, and essentialism has an experimental tradition, not that old, by the way, in naïve biology. The way that children reason about the world, there's a lot of good evidence to suggest that there are domains of knowledge: physical, reasoning about the physical world; reasoning about the living world, the biological one; and reasoning about the psychological world. Those three domains are the physics, the biology, and the psychology, and are deemed to cover the majority of what we do when we're thinking about concepts.

In the biological world, people like Gelman have argued that children infer an invisible dimension. When they're making categorical decisions about why dogs are different from cats, for example, they go over and beyond the outward appearances, and infer that there must be some internal property that makes a dog a dog. Irrespective of changing its outward appearance or if you raise it with a litter of kittens, it will still turn out and grow up into a dog. So they kind of understand there's something over and beyond the physical aspect of it. Well, there is, it's DNA, but no four-year-old knows explicitly about DNA. But they do have this intuition that there is this essential property. Essentialism in the research field, in developmental psychology, started off in biology. But I was interested in essentialism basically almost contaminating to different domains for objects, the way that we treat objects as irreplaceable. This is the issue of authenticity. The authenticity of objects is starting to get into the boundary of what, of an essential property, makes something irreplaceable.

This is work I've done with Paul Bloom. We initially started looking at sentimental objects, the emergence of this bizarre behavior that you find in children in the West. They form these emotional attachments to blankets and teddy bears, and it initially starts off as an associative learning type of situation where they need to self-soothe, because in the West we typically separate children, for sleeping purposes, between one and two years of age. In the Far East they don't; they keep children well into middle childhood, so they don't have as much attachment object behavior. It's common—about three out of four children start off with this sort of attachment to particular objects and then it dissipates and disappears.

What Paul and I are interested in is whether or not it was the physical properties of the object or if there was something about the identity or the authenticity of the object which is important. We embarked on a series of studies where we convinced children we had a duplicating machine, and basically we used conjuring tricks to convince the child that we could duplicate any physical object. We have these boxes which looked very scientific, with wires and lights, and we place an object in one, and activate it, and after a few seconds the other box would appear to start up by itself and you open it up and you see you've got two identical objects. The child spontaneously said, "Oh, it's like a copying machine." It's like a photocopier for objects, if you like. Once they're in the mind-set that this thing can copy, we then test what you can get away with. They're quite happy to have their objects, their toys copied, but when it comes to a sentimental object like a blanket or a teddy bear, then they're much more resistant to accepting the duplicate. Now, of course, we can't actually copy the things, but they think that when the machine starts up there will be another duplicated object. That's how we test their intuitions and inferences on it.

On the basis of that, we've started to—just completed—a set of studies looking at copying real things, living things, like hamsters. Because the question is, would a child think that a machine that duplicates objects could duplicate minds? We're getting into the issue about minds being separate or a product of bodies. The work is still under review, so I can't say too much about it, but you can see the obvious link to the philosophical issues about mind-body dualism.

Also, we're getting into the territory of authenticity and identity. There are some fairly old philosophical issues about what confers identity and uniqueness, and these are the principles, quiddity and haecceity. I hadn't even heard of these issues until I started to research into it, and it turns out these obscure terms come from the philosopher Duns Scotus. Quiddity is the invisible properties, the essence shared by members of a group, so that would be the "dogginess" of all dogs. But the haecceity is the unique property of the individual, so that would be Fido's haecceity or Fido's essence, which makes Fido distinct to another dog, for example.

These are not real properties. These are psychological constructs, and I think the reason that people generate these constructs is that when they invest some emotional time or effort into an object, or it has some significance toward them, then they imbue it with this property, which makes it irreplaceable—you can't duplicate it. In effect, it becomes sacred, and so I think that sacred objects, which exist across various religions, also have this notion of them being unique. You can't duplicate and you can't corrupt them. They have this property that is indivisible. I think essentialism is pervasive in our attitudes toward objects, but it's also there in our attitudes to valuation.

Paul has done more work on this about works of art, what makes a masterpiece unique, and it turns out that our intuitions are often guided by this sense that there's an additional property over and beyond the actual sort of physical aspects of works of art. But I think you can also see it operating in other things.

For example, a lot of marketing. When you're selling products, whether through design or just by sheer experience, what I think a lot of advertisers have realized is that certain properties, if you emphasize the essential quality of the object, this confers the notion of quality. Coors beer, from what I understand, was going to relocate their brewery, but then there was this concern that they wouldn't have the original source of the water. If you start to think about a lot of luxury products, there's this notion of craftsmanship, of something of the maker going into the product. So I also think this is a principle or way of thinking that affects and influences how we think about genetic modification. Tinkering with the essence or the natural properties of things—people don't like the idea of it. It just seems to violate what are core integrity principles.

I see essentialism everywhere I look now. It just seems to be pervasive. It's one of these ways of seeing the world. People say, Oh, it might just be association, but association strikes me as inadequate as an explanation as to why some things seem to have this property. Of course, Paul Rozin's work on moral contamination and contagion, again, I think speaks to this idea there are things which can contaminate you with evil—for example, just by wearing a killer's cardigan, things like that. So as I say, I see it everywhere. I'm hoping to continue that kind of work. So that's an example of a philosophical kind of question, or certainly a philosophical domain that I think does lend itself to the empirical studies and, who knows, that might actually turn out to have some application as far as marketers are concerned.

There are actual real colleagues, Susan Gelman and Paul Bloom, and George Newman is the psychologist who's now in the economics school at Yale. We're planning on doing some research looking at advertising and looking at what aspects of it conform to this essentialist principle. I've done some work with my graduate students showing that children have intuitions about ownership of objects. If somebody has put some craftsmanship into it—for example, if you've got some clay and I've got some clay and I take your clay and make something, and if I said, "Who owns it?" an adult would try and probably evaluate that ownership on the basis of how much effort went into it, and also who owned the original material. But young children spontaneously say, well, irrespective of who owned the material, it's actually the person who changed and transformed it. The ownership transfers to the person of the craftsmanship.

That's interesting because it gets to whole questions about intellectual property. Who had the original idea? Is there any such thing as an original idea? To what extent does modification of the idea transfer the ownership of the intellectual property? So, again, there are all these kinds of issues that are, we have intuitions about, and some of them are culturally defined, but I think there's a developmental process. I think that children start off with some basic ideas that then become modified and changed through culture. The detractors of the whole essentialist position are likely to be those from the associative schools who argue that a lot of behaviors can be explained away by simple mechanisms of learning and reinforcement, but that tends to be the older school. Nobody comes to mind about that.

The self is something that is central to a lot of psychological questions, and, in fact, a lot of psychologists have difficulty describing their work without positing the notion of a self. It's such a common daily, profound, indivisible experience for most of us. Some people do manage to achieve states of divided self or *anatta*, no self—they're really skilled Buddhists. But for the majority of us the self is a very compulsive experience. I happen to think it's an illusion, and certainly the neuroscience seems to support that contention. Simply from the logical positions that it's very difficult to, without avoiding some degree of infinite regress, to say a starting point, the trail of thought, just the fractionation of the mind, when we see this happening in neurological conditions. The famous split-brain studies showing that actually we're not integrated entities inside our head, rather we're the output of a multitude of unconscious processes.

I happen to think the self is a narrative, and I use the self and the division that was drawn by William James, which is the "I" (the experience of conscious self) and the "me" (which is personal identity, how you would describe yourself in terms of where are you from and everything that makes you up in your predilections and your wishes for the future). Both the "I," who is sentient of the "me," and the "me," which is a story of who you are, I think are stories. They're constructs and narratives. I mean that in a sense that a story is a reduction, or at least a coherent framework that has some causal kind of coherence.

When I go out and give public lectures I like to illustrate the weaknesses of the "I" by using visual illusions of the most common examples. But there are other kinds of illusions that you can introduce, which just reveal to people how their conscious experience is actually really just a fraction of what's really going on. It certainly is not a true reflection of all mechanisms that are generating. Visual illusions are very obvious in that. The thing about the visual illusion effects is that even when they're explained to you, you can't but help see them, so that's interesting. You can't divorce yourself from the mechanisms that are creating the illusion and the mind that's experienced in the illusion.

The sense of personal identity, this is where we've been doing experimental work showing the importance that we place upon episodic memories, autobiographical memories. In our duplication studies, for example, children are quite willing to accept that you could copy a hamster with all its physical properties that you can't necessarily see, but what you can't copy very easily are the episodic memories that one hamster has had.

This actually resonates with the ideas of John Locke, the philosopher, who also argued that personal identity was really dependent on the autobiographical or episodic memories, and you are the sum of your memories, which, of course, is something that fractionates and fragments in various forms of dementia. As the person loses the capacity to retrieve memories, or these memories become distorted, then the identity of the person, the personality, can be changed, among other things. But certainly the memories are very important.

As we all know, memory is notoriously fallible. It's not cast in stone. It's not something that is stable. It's constantly reshaping

Bruce Hood

itself. So the fact that we have a multitude of unconscious processes that are generating this coherence of consciousness, which is the "I" experience, and the truth that our memories are very selective and ultimately corruptible—we tend to remember things that fit with our general characterization of what our self is. We tend to ignore all the information that is inconsistent. We have all these attribution biases. We have cognitive dissonance. The very thing psychology keeps telling us, that we have all these unconscious mechanisms that reframe information, to fit with a coherent story, then both the "I" and the "me," to all intents and purposes, are generated narratives.

The illusions I talk about often are this sense that there is an integrated individual, with a veridical notion of past. And there's nothing at the center. We're the product of the emergent property, I would argue, of the multitude of these processes that generate us.

I use the word *illusion* as opposed to delusion. Delusion implies mental illness, to some extent, and illusion, we're quite happy to accept that we're experiencing illusions, and for me the word *illusion* really does mean that it's an experience that is not what it seems. I'm not denying that there is an experience. We all have this experience, and what's more, you can't escape it easily. I think it's more acceptable to call it an illusion, whereas there's a derogatory nature of calling something a delusion. I suspect there's probably a technical difference that ought to do with mental illness, but no, I think we're all perfectly normally experiencing this illusion.

Oliver Sacks has famously written about various case studies of patients who seem so bizarre, people who have various forms of perceptual anomalies, they mistake their wife for a hat, or there are patients who can't help but copy everything they see. I think that in many instances, because the self is so core to our normal behavior, having an understanding of that self is this constructive process. I think if this was something that clinicians were familiar with, then I think that would make a lot of sense.

In fact, it's not only in clinical practice, I think in a lot of things. I think neuroethics is a very interesting field. I've got another colleague, David Eagleman—he's very interested in these ideas. The culpability, responsibility. We premise our legal systems on this notion that there is an individual who is to be held accountable. Now, I'm not suggesting that we abandon that, and I'm not sure what you would put in its place, but I think we can all recognize that there are certain situations where we find it very difficult to attribute blame to someone. For example, famously, Charles Whitman, the Texan sniper. When they had the autopsy, they discovered a very sizable tumor in a region of the brain which could have very much influenced his ability to control his rage. I'm not suggesting every mass murderer has inoperable tumors in their brain, but it's conceivable that there will be—with our increasing knowledge of how the brain operates, and our ability to understand it—it's conceivable there will be more situations where the lawyers will be looking to put the blame on some biological abnormality.

Where is the line to be drawn? I think that's a very tough one to deal with. It's a problem that's not going to go away. It's something that we're going to continually face as we start to learn more about the genetics of aggression.

There's a lot of interest in this thing called the warrior gene. To what extent is this a gene that predisposes you to violence? Or do you need the interaction between the gene and the abusive childhood in order to get this kind of profile? So it's not just clinicians, it's actually just about every realm of human activity where you posit the existence of a self and individuals, and responsibility. Then it will reframe the way you think about things. Just the way that we heap blame and praise, the flip side of blaming people is that we praise individuals. But it could be, in a sense, a multitude of factors that have led them to be successful. I think that it's a pervasive notion. Whether or not we actually change the way we do anything, I'm not so sure, because I think it would be really hard to live our lives dealing with nonindividuals, trying to deal with multitude and the history that everyone brings to the table. There's a good reason why we have this experience of the self. It's a very sort of succinct and economical way of interacting with each other. We deal with individuals. We fall in love with individuals, not multitudes of past experiences and aspects of hidden agendas—we just pick them out.

I think it's unlikely we're ever going to get rid of the Great Selfini, as Dan Wegner has called this character. He's revealed this character, the Great Selfini. I love that. But even when you know that it's likely to be this constructed mental experience, it doesn't give you any insight, it doesn't stop you treating and thinking like that.

Laurie Santos and I, we go back quite a bit. Laurie was actually a brilliant undergraduate at Harvard when I was visiting at MIT, and we got interested in naïve physics. I'd discovered an unusual error that children make, namely that they assume that when you drop something it always falls in a straight line. We used to test this with an arrangement of tubes, and you drop an object down it, and children typically would search directly below, over and over and over again, so they found it very difficult to overcome this tendency to think that objects fall straight down. I actually thought it was not so much an error, I thought it was a very good, naïve theory to have, because if you do drop something, more often than not it's going to be directly below where you released it. I thought it was a very reasonable way to think about the world. They don't start off with that, so they have to acquire this probably through experience.

The question was, what about nonhuman primates? At the time Laurie was working in the labs at Harvard that were doing a lot of these comparisons between the human developmental studies and various nonhuman primates. So that's where we first met up. Then when I moved there Laurie was one of my students and, frankly, was much better at understanding the concepts than I was at actually teaching them. So from a very early point I knew that she was quite exceptional. Laurie has gone on to establish herself with a remarkable lab.

The questions of interest that we work on together are, we've done some work on object identity looking at Michotte's original observations of the tunnel illusion. Michotte first reported that if you observe an object entering a tunnel and a different object exiting at the right time, then very often you have this impression that it's the same object that is transformed. It's called the tunnel illusion. She studied this originally with rhesus monkeys, and in a twist of the usual normal order of the way things get done, we actually saw the animal studies and then developed a child paradigm for that, so we tested children with a tunnel task illusion, so they'd see an object go into one tunnel and then reemerge, and then exit into a second hiding place. Now, if they really did believe it was just one object transforming, if they had to go to find it, they'd look at the end of the sequence. But if they understood that in fact there must have been two objects present, they would know to look in the first tunnel and the second tunnel. We use this as a way of testing what their perceptual experiences were. We've been writing that together.

More recently, Laurie's been working on the endowment effect, which, of course, is now one of the most famous effects from be-

Bruce Hood

havioral economics, namely that you overvalue something that you perceive is in your ownership. Laurie has been looking at that with her capuchins. I have another graduate student, we've been looking at it in a variety of primates—so we've done the gorillas, the orangs, the chimps as well. Actually a couple of labs have been doing this.

What I think is a good consensus of opinion is that you don't find the endowment effect in nonhuman primates in anything other than food. They'll show the effects of ownership and overvaluing the food that's in their possession, but they won't do it for tools and they won't do it for tools to retrieve food. So this thing for objects, I think, is a human attribute. I keep tying this back to my issues about why certain objects are overvalued, and I happen to believe, like James again, that objects are part of the extended sense of self. We surround ourselves with objects. We place a lot of value on objects that we think are representative of our self.

That's an area which might tick off those boxes I was mentioning early on, namely that it has application. It's actually theoretically quite interesting as well. It intrigues me. We're the only species on this planet that invests a lot of time and evaluation through our objects, and this has been something that has been with us for a very, very long time.

Think of some of the early artifacts. The difficulty would have been to make these artifacts, the time invested in these things, means that from a very early point in our civilization, or before civilization, I think the earliest pieces are probably about 90,000 years old. There are certainly older things that are tools, but pieces of artwork, about 90,000 years old. So it's been with us a long time. And yes, some of them are obviously sacred objects, power of religious purposes and so forth. But outside of that, there's still this sense of having materials or things that we value, and that intrigues me in so many ways. And I don't think it's necessarily universal as well. It's been around a lot, but the endowment effect, for example, is not found everywhere. There's some intriguing work coming out of Africa.

The endowment effect is this rather intriguing idea that we will spontaneously overvalue an object as soon as we believe it's in our possession. We don't actually have to have it physically, just bidding on something—as soon as you make your connection to an object, then you value it more, you'll actually remember more about it, you'll remember objects that you think are in your possession in comparison to someone else's. It gets a whole sense of attribution and value associated with it, which is one of the reasons that people never get the asking price for the things that they're trying to sell, they always think their objects are worth more than other people are willing to pay for them.

There was the first experimental demonstration by Richard Thaler and Danny Kahneman, and the early behavioral economics, which was this demonstration that if you just give people coffee cups, students, coffee cups, and then you ask them to sell them, they always ask for more than what someone's willing to pay for it. It turns out it's not just coffee cups, it's wine, it's chocolate, it's anything, basically. There's been quite a bit of work done on the endowment effect now. As I say, it's been looked at in different species, and the brain mechanisms of having to sell something at a lower price, like loss aversion, it's seen as quite painful, triggers the same pain centers, if you think you're going to lose out on a deal.

What is it about the objects that give us this self-evaluated sense? Well, I think William James spoke of this, again, on the way that we use objects to extend our self. Russell Belk is a marketing psychologist. He has also talked about the extended self in terms of objects. As I say, this is something that I think marketers know, in that they create certain quality brands that are perceived to signal to others how good your social status is.

It's something in us, but it may not be universal because there are tribes, there are some recent reports from nomadic tribes in central Africa, who don't seem to have this sense of ownership. It might be a reflection more of the fact that a lot of this work has been done in the West where we're very individualistic, and of course individualism almost creates a lot of endowment ideas and certainly supports the endowment, materialism that we see. But this is an area I'd like to do more work with, because we have not found any evidence of the endowment effect in children below five, six years of age. I'm interested: is this something that just emerges spontaneously? I suspect not. I suspect this is something that culture is definitely shaping. That's my hunch, so that's an empirical question I need to pick apart.

Another line of research I've been working on in the past five years . . . this was a little bit like putting the cart before the horse, so I put forward an idea, it wasn't entirely original. It was a combination of ideas of others, most notably Pascal Boyer. Paul Bloom, to some extent, had been thinking something similar. A bunch of us were interested in why religion was around. I didn't want to specifically focus on religion. I wanted to get to the more general point about belief, because it was my hunch that even a lot of atheists or self-stated atheists or agnostics still nevertheless entertained beliefs which were pretty irrational. I wasn't meaning irrational in a kind of behavioral economics type of way. I meant irrational in that there were these implicit views that would violate the natural laws as we thought about them. Violations of the natural laws I see as being supernatural. That's what makes them supernatural. I felt that this was an area worth looking at. They'd been looked at 50, 60 years ago very much in the behaviorist association tradition.

B. F. Skinner famously wrote a paper on the superstitious behavior of pigeons, and he argued that if you simply set up a reinforcement schedule at a random kind of interval, pigeons will adopt typical patterns that they think are somehow related to the reward, and then you could shape irrational superstitious behaviors. Now that work has turned out to be a bit dubious, and I'm not sure that stood the test of time. But in terms of people's rituals and routines it's quite clear, and I know them in myself. There are these things that we do that are familiar, and we get a little bit irritated if we don't get to do them, so we do, most of us, entertain some degree of superstitious behavior.

At the time there was a lot of interest in religion and a lot of the hoo-ha about The God Delusion, and I felt that maybe we just need to redress this idea that it's all to do with indoctrination, because I couldn't believe that the whole edifice of this kind of belief system was purely indoctrination. I'm not saying there's not indoctrination, and clearly religions are culturally transmitted. You're not born to be Jewish or born to be Christian. But what I think religions do is capitalize on a lot of inclinations that children have. Then I entered into a series of work, and my particular interest was this idea of essentialism and sacred objects and moral contamination.

We took a lot of the work that Paul Rozin had done, talking about things like killers' cardigans, and we started to see whether there was any empirical measures of transfer. For example, would you find yourself wanting to wash your hands more? Would you find priming effects for words which were related to good and evil, based on whether you had touched the object or not? For me there had to be this issue of physical contact. It struck me that this was why it wasn't a pure association mechanism. It was actually something to do with the belief, a naïve belief, there was some biological entity that can somehow—moral contamination can transfer.

We started to look at actually not children now, but looking at adults, because doing this sort of work with children is very difficult and probably somewhat controversial. But the whole area of research is premised on this idea that there are intuitive ways of seeing the world. Sometimes this is referred to as system one and system two, or automatic and control. It reappears in a variety of psychological contexts. I just think about it as these unconscious, rapid systems that are triggered automatically. I think their origins are in children. While you can educate people with a kind of slower system two, if you like, you never eradicate the intuitive ways of seeing the world because they were never taught in the first place. They're always there. I suppose if you want to ask me if there is any kind of thing that you can have as a theory that you haven't yet proven, it's the idea that I don't think you ever throw away any belief system or any ideas that have been derived through these unconscious intuitive processes. You can supersede them, you can overwrite them, but they never go away, and they will reemerge under the right contexts. If you put people through stressful situations or you overload them, you can see the reemergence of these kinds of ways of thinking. The empirical evidence seems to be supporting that. They've got wrinkles in their brains. They're never going to go away. You can try to override them, but they're always there and they will reappear under the right circumstances, which is why you see the reemergence under stress of a lot of irrational thinking.

For example, teleological explanations, the idea that everything is made for a purpose or a function, is a natural way to see the world. This is Deb Kelemen's work. You will find that people who considered themselves fairly rational and well educated will, nevertheless, default back to teleological explanations if you put them under a stressful, timed kind of situation. So it's a way of seeing the world that is never eradicated. I think that's going to be a general principle, in the same way that a reflex, if you think about reflexes, that's an unlearned behavioral response. You're born with a whole set of reflexes. Many of them disappear, but they never entirely go away. They typically become reintegrated into more complex behaviors, but if someone goes into a coma, you can see the reflexes reemerging.

What we think is going on is that in the course of development, these very automatic behaviors become controlled by top-down processes from the cortex, all these higher-order systems that are regulating and controlling and suppressing, trying to keep these things under wraps. But when the cortex is put out of action through a coma or head injury, then you can see many of these things reemerging again. I don't see why there should be any point of departure from a motor system to a perceptual system, to a cognitive system, because they're all basically patterns of neural firing in the brain, and so I don't see why it can't be the case that if concepts are derived through these processes, they could remain dormant and latent as well.

One of the things that has been fascinating me is the extent to which we can talk about the hierarchy of representations in the brain. Representations are literally re-presentations. That's the language of the brain, that's the mode of thinking in the brain, it's representation. It's more than likely, in fact, it's most likely that there is already representation wired into the brain. If you think about the sensory systems, the array of the eye, for example, is already laid out in a topographical representation of the external world, to which it has not yet been exposed. What happens is that

Bruce Hood

this is general layout, arrangements that become fine-tuned. We know of a lot of work to show that the arrangements of the sensory mechanisms do have a spatial arrangement, so that's not learned in any sense. But these can become changed through experiences, and that's why the early work of Hubel and Wiesel about the effects of abnormal environments showed that the general pattern could be distorted, but the pattern was already in place in the first place.

When you start to move beyond sensory into perceptual systems and then into cognitive systems, that's when you get into theoretical arguments and the gloves come off. There are some people who argue that it has to be the case that there are certain primitives built into the conceptual systems. I'm talking about the work of, most notably, Elizabeth Spelke.

There certainly seems to be a lot of perceptual ability in newborns in terms of constancies, noticing invariant aspects of the physical world. I don't think I have a problem with any of that, but I suppose this is where the debates go. To what extent are concepts built in? There are some people who could spend the next five hours telling you about that, but I'm not prepared to try to do so.

Shame, to me, is interesting. I've been to Japan a couple of times. I'm not an expert in the cultural variation of cognition, but clearly shame is a major factor in motivation, or avoidance of shame, in Eastern cultures. I think it reflects the sense of self-worth and value in Eastern culture. It is very much a collective notion that they place a lot of emphasis on not letting the team down. I believe they even have a special word for that aspect or experience of shame that we don't have. That doesn't mean that it's a concept that we can never entertain, but it does suggest that in the East this is something that is at least recognized as a major factor of identity.

Children don't necessarily feel shame. I don't think they've got a sense of self until well into their second year. They have the "I," they have the notion of being, of having control. They will experience the willingness to move their arms, and I'm sure they make that connection very quickly, so they have this sense of self, in that "I" notion, but I don't think they've got personal identity, and that's one of the reasons that they don't have much, or very few of us have much, memory of our earlier times. Our episodic memories are very fragmented, sensory events. But from about two to three years on they start to get a sense of who they are. Knowing who you are means becoming integrated into your social environment, and part of becoming integrated into your social environment means acquiring a sense of shame. Below two or three years of age, I don't think many children have a notion of shame. But from then on, as they have to become members of the social tribe, then they have to be made aware of the consequences of being antisocial or doing things that are not what's expected of them. I think that's probably late in the acquisition.

Let me tell you a little bit about what's happening. As part of this transition in my career, which is to do more public engagement, I've noticed a couple of what I think are growing trends. For example, there is a growing skeptical movement which originated as a grassroots movement—just, basically, nonscientist members of the public have become very interested in scientists and speaking to scientists and engaging the scientists. There's one organization called Skeptics in the Pub, and it's a nonprofit organization, there's no money involved, we do it all for the love of it, and we'll turn up and give maybe an hour's presentation in the back room of a pub. That started off when I started doing this about five years ago. I think there were maybe three or four, and now I think there are 30 up and down the country. There are these grassroots movements to try to integrate

Bruce Hood

with science. But I don't think that—certainly my field, and I'm not sure this is the case for other areas of science, but behavioral science has had a very poor reputation in this country. I think that's partly historical, and I think that's partly traced back to the dominance of the psychoanalytic movement in America, and its ridicule in this country.

Psychology has always been regarded as not a science. I have a personal agenda to try to reinforce in people that psychology is actually a really important science and remind them that the last pages of *On the Origin of Species*, this is the prediction made by Charles Darwin: that the next leap forward in understanding the human condition would involve psychology. I keep reminding people that they shouldn't assume that when someone goes on television and tells you how you should try to attract somebody or choose someone for a job, that that is the full extent of psychology. But that, unfortunately, is what the public has been fed as a diet of psychological work. Clearly, as soon as you say you're a cognitive scientist, they don't know what that means, so you can kind of redescribe yourself, but then you start addressing issues about the philosophical problems, behavioral economics. Then it becomes more palatable or more acceptable as a legitimate science.

I feel that there is a shift toward engagement with scientists, and I'm up for that, and I want to do that, and I have a personal agenda to try to make behavioral science much more on a par with the other sciences. It's still very popular with the students in the universities because it's a lot cheaper than undertaking other types of sciences, which are that much more expensive. But I think in the States, my impression when I was there was that we had a much greater social standing in the general public, and I think we're behind the United States in that way and we could do with a sort of making up some of that lost ground.

9

Testosterone on My Mind and in My Brain

Simon Baron-Cohen

Psychologist; Professor of Developmental Psychopathology and Director of the Autism Research Centre, Cambridge University; Fellow of Trinity College, Cambridge; author, *The Essential Difference* and *The Science of Evil*.

INTRODUCTION by John Brockman

"I thoroughly enjoyed the evening last week," emailed Brian Eno. "A lot of interesting people got connected together and everyone told me they enjoyed themselves."

He was referring the first-ever London Edge Reality Club meeting, featuring a presentation by Cambridge research psychologist Simon Baron-Cohen, which Eno hosted at his studio in Notting Hill before an assembled group that included artists, curators, museum directors, writers, playwrights, scientists in fields such as biology, math, psychology, and zoology, and the editors and correspondents of Nature, The Economist, Wired, *and* The Guardian.

Baron-Cohen held forth before this diverse group on his latest research on the properties and effects of the hormone testosterone, while showing its relevance to his earlier research in sex differences and autism.

The variety of people at the meeting reflected the singular place in the London cultural scene occupied by our host, an artist, "nonmusician," composer, columnist, and record producer (U2, Coldplay, Talking Heads, Paul Simon). "Be happy to make it available whenever you need it." Short answer: an emphatic yes. But first, we need to check with Larry and Ser-

gey to make sure that Google's search algorithms and infrastructure are robust enough to handle the combustible combination of Eno's dual relationships to "Edge" (the guitarist for U2) and "Edge" (edge.org).

What I want to talk about tonight is this very specific hormone, testosterone. Our lab has been doing a lot of research to understand what this hormone does and, in particular, to test whether it plays any role in how the mind and the brain develop.

Before I get to that point, I'll say a few words by way of background about typical sex differences, because that's the cradle out of which this new research comes. Many of you know that the topic of sex differences in psychology is fraught with controversy. It's an area where people, for many decades, didn't really want to enter because of the risks of political incorrectness, and of being misunderstood.

Perhaps of all of the areas in psychology where people do research, the field of sex differences was kind of off-limits. It was taboo, and that was partly because people believed that anyone who tried to do research into whether boys and girls, on average, differ must have some sexist agenda. And so for that reason a lot of scientists just wouldn't even touch it.

By 2003, I was beginning to sense that that political climate was changing, that it was a time when people could ask the question—Do boys and girls differ? Do men and women differ?—without fear of being accused of some kind of sexist agenda, but in a more open-minded way.

First of all, I started off looking at neuroanatomy, to look at what the neuroscience is telling us about the male and female brain. If you just take groups of girls and groups of boys and, for example, put them into MRI scanners to look at the brain, you do see differences on average. Take the idea that the sexes are identical from the neck upward, even if they are very clearly different from the neck downward: the neuroscience is telling us that that is just a myth, that there are differences, even in terms of brain volume and the number of connections between nerve cells in the brain at the structure of the brain, on average, between males and females.

I say this carefully because it's still a field which is prone to misunderstanding and misinterpretation, but just giving you some of the examples of findings that have come out of the neuroscience of sex differences, you find that the male brain, on average, is about eight percent larger than the female brain. We're talking about a volumetric difference. It doesn't necessarily mean anything, but that's just a finding that's consistently found. You find that difference from the earliest point you can put babies into the scanner, so some of the studies are at two weeks old in terms of infants.

You also find that if you look at postmortem tissue, looking at the human brain in terms of postmortem tissue, that the male brain has more connections, more synapses between nerve cells. It's about a 30 percent difference on average between males and females. These differences are there.

The second big difference between males and females is about how much gray matter and white matter we see in the brain: that males have more gray matter and more white matter on average than the female brain does. White matter, just to be succinct, is mostly about connections between different parts of the brain. The gray matter is more about the cell bodies in the brain. But those differences exist. Then when you probe a bit further, you find that there are differences between the male and female brain in different lobes, the frontal lobe, the temporal lobe, in terms of how much gray and white matter there is.

Simon Baron-Cohen

You can also dissect the brain to look at specific regions. Some of you will have had heard of regions like the amygdala, which people think of as a sort of emotion center, that tends to be larger in the male brain than the female brain, again, on average. There's another region that shows the opposite pattern, larger in females than males: the planum temporale, an area involved in language. These structural differences exist, and I started by looking at these differences in terms of neuroanatomy, just because I thought at least those are differences that are rooted in biology, and there might be less scope for disagreement about basic differences.

I've talked a little bit about neuroanatomy, but in terms of psychology, there are also sex differences that are reported. On average, females are developing empathy at a faster rate than males. I keep using that word "on average" because none of these findings apply to all females or all males. You simply see differences emerge when you compare groups of males and females. Empathy seems to be developing faster in girls, and in contrast, in boys there seems to be a stronger drive to systemize. I use that word "systemizing," which is all about trying to figure out how systems work, becoming fascinated with systems. And systems could take a variety of different forms. It could be a mechanical system, like a computer; it could be a natural system, like the weather; it could be an abstract system, like mathematics; but boys seem to have a stronger interest in systematic information. I was contrasting these two very different psychological processes, empathy and systemizing. And that's about as far as I went, and that was now some 11 years ago.

Since then my lab has wanted to try to understand where these sex differences come from, and now I'm fast-forwarding to tell you about the work that we're doing on testosterone. I'm very interested in this molecule, partly because males produce more of it than females, and partly because there's a long tradition of animal research which shows that this hormone may masculinize the brain, but there's very little work on this hormone in humans.

Before I tell you about the testosterone work, I should just mention one somewhat controversial study that we published back in the year 2000, and that was a study of newborn babies. What we wanted to do was establish whether the sex differences that you find in behavior, or in the mind, in human beings were purely the result of culture and purely the result of postnatal experience, or whether biology might also be contributing to those differences.

We did a study that we call the Newborn Baby Study, where we studied just over 100 babies aged 24 hours old. This was a study conducted in Cambridge. The babies had just popped out of the womb. The mothers had signed a consent form, saying that they were happy for their baby to take part in research. We would have loved to study the babies as soon as they came out, but the obstetricians asked us, out of respect for mother and baby, if we would wait 24 hours, which we were happy to do; and then when the babies were settled and the mother was settled, we presented these babies with two objects: a human face or a mechanical mobile suspended above the crib. So, two very different kinds of objects, one mechanical and one animate and human. And we looked to see whether babies, aged one day old, looked longer at the human face, a social stimulus, or looked longer at the mechanical object.

The objects were presented one at a time, and we counterbalanced—that's to say we varied the order—whether the baby saw the face first or the mechanical object first. What we found, just cutting to the chase, was that if we compared babies in terms of looking longer at a social stimulus or looking longer at a mechanical stimulus, more boys seemed to look longer at the mechanical stimulus, and more girls seemed to look longer at the

Simon Baron-Cohen

social stimulus, the face. We were finding a difference that was there as early as 24 hours old; and that, to me, was pointing to biology being a contributing factor to sex differences.

People who really want to argue that all sex differences that you find in the mind and in behavior can be explained by cultural factors might still have argued that in those 24 hours, mother and baby or father and baby had been interacting, that parents may have been somehow shaping their sons and daughters to have different patterns of interest. That is just possible, because in 24 hours, people would argue, there's plenty of room for experience. But it's equally plausible that what we're seeing is babies arriving on day one with slightly different patterns of interest: with girls on average being more oriented toward people, being more inquisitive about people, and boys, again on average, being slightly more oriented in their attention toward the physical environment, and patterns in the physical world.

Now we're up to testosterone. We picked testosterone, this molecule, just because we were looking for a candidate for biological explanation or a mechanism for these sex differences we were observing. And the animal research was pointing to the fact that before birth there's a surge in the production of testosterone. Testosterone is suddenly produced at very high quantities, and then it drops off again around birth. And the animal researchers were arguing that this surge in the production of testosterone, that the fetus is producing a lot of testosterone during fetal life, because it has some permanent and organizing effect on brain development. And they were able to show that in rats.

You can do experimental manipulations of this hormone in rats that would be completely unethical in humans: you can either deprive the animal of its testosterone, for example, by castration at birth (one source of it in males is the testes); or you can inject extra testosterone, for example, into a female rat at birth, and look at the effects of brain and behavior. And that experimental evidence was pointing to long-term "organizational" effects on brain development.

Just to take an example: if you have a female rat that's given extra testosterone either during pregnancy or at birth, and then look at that female rat's behavior postnatally, her behavior is much more like a typical male rat. Ways that you can test this include, for example, letting the rats run through mazes, finding their way through a maze. If you take male and female rats, usually the male rats will get through the maze faster, learning a spatial route more quickly; but female rats that have been given extra testosterone are much more like typical male rats, so their behavior has been changed. And then if you look at the postmortem rat brain, the female rats that were given extra testosterone, in terms of their neuroanatomy, the female rat's brain looks much more like a typical male rat's brain.

There was a hypothesis from animal research that testosterone was this special molecule that masculinizes the brain, and it was looking clearer and clearer in other species, but no one had found a way to demonstrate it in humans. What we've been doing over the last decade is to try to find a way to test it in humans. And the way that we settled on was to look at women in pregnancy who were having a procedure called amniocentesis. Many of you will have heard of this: this is where a pregnant woman has a long needle introduced into the womb, the baby is in the womb surrounded by amniotic fluid, and the needle goes in as part of a medical, clinical procedure, to take some of that fluid. That's called amniocentesis, where you take some of the amniotic fluid that surrounds the baby. Usually this is done because there's some suspicion that the baby might have Down syndrome, and the doctor wants to analyze the amniotic fluid for chromosomal abnormalities, as a test of whether the baby will indeed go on to have Down syndrome.

These women were having this procedure, and the research had been telling us that if you wanted to look at hormones like testosterone, in terms of their potential influence on the human brain, you'd need to find a way of measuring it during pregnancy. We put these two things together and asked these women, if you're having an amniocentesis anyway, can we take some of the fluid, with your consent, and analyze it for testosterone levels that you find in the fluid? This is when the baby is sometime between 12 and 19 weeks of pregnancy. It's a little window when amniocentesis takes place. If you think of pregnancy as 40 weeks normally, nine months, roughly 40 weeks, this is at the end of the first trimester, the first third of pregnancy, and just going into the second trimester, the middle of pregnancy.

We were getting this opportunity to measure hormones, but particularly testosterone, whilst the baby was still growing, at a point when the brain is developing very rapidly. And then the design of our research was that we would then store the fluid in our deep freeze, wait for the baby to be born, and follow up the babies after they were born to see if there was any correlation between their testosterone levels in the womb and how they turned out as children.

I should say one other thing, which is that this was premised on the idea that children are more different to each other than they are similar. Those of you who are parents know this very well: that you might have two or three kids and despite the fact that they've grown up in the same home, they've got the same parents, they seem to have very different personalities, very different patterns of interest, and very different rates of development. Part of the challenge, scientifically, was also to try to understand this variability in development, individual differences in development, and we were anchoring our research by looking at one factor, testosterone, which varies, with some people being very low in testosterone, other people being average or very high in testosterone, and seeing whether that scale of individual differences had anything to do with individual differences we see postnatally—for example, in rates of language development, in how sociable children are, and on other dimensions.

The study was that the woman had her amniocentesis while she was pregnant, we then waited for the baby to be born, and we've invited these babies in pretty much every year since their birth. We started off with about 500 of these babies—that's the cohort—where their testosterone levels are known, and we've been following them and they're now about 12 years old. So it's a story that's been unfolding whilst we've been able to measure the behavior as they grow, and see whether it has anything to do with their testosterone prenatally.

At their first birthday, we looked at eye contact. And I was particularly interested in this, because my main area of research is autism, and children with autism make very little eye contact. Their eye contact is at the extreme, showing very little interest in faces. But we'd already heard from that newborn baby study that there seems to be, on average, a sex difference in how interested people are in faces, and making eye contact, with girls being, on average, more interested in faces than boys, but there's a whole spectrum of individual differences. And what we found was that the higher the baby's prenatal testosterone, the less eye contact they made at their first birthday. That was simply measured by inviting the toddler into our lab, videotaping them, and then later coding those videotapes for how many times the baby looked up at their mother's face. That was at the first birthday.

At the second birthday we looked at language development. We got parents to fill in a checklist of how many words does your child know, and how many words can your child produce. We were looking at the size of children's vocabularies. What we found, which was quite striking to us, was that by two years old, there were some children who had very small vocabularies, only about 10 or 20 words, they were kind of at the low edge of normal development; and there were some kids who were really chatty and had 600 words. So the size of the differences in vocabulary was immense. And then we could look back at their testosterone levels. And once again, we found a significant correlation: that the higher the child's prenatal testosterone, the smaller their vocabulary at two years old. So this same hormone seemed to be related not only to patterns of social interest—whether you look at faces—but also to communication—how talkative you are and your rate of language development.

I don't want to go through all the steps, but we've also looked, when they were four years old, at empathy, finding that prenatal testosterone is negatively correlated with empathy. So again, it's the same pattern we were just hearing about: that if you were higher in testosterone during the pregnancy, it meant that you were slower to develop empathy as a four-year-old. And again, there are different ways that you can measure that. You can ask parents to fill in questionnaires about their child's empathy. You can actually get the child to take various empathy tests, or you can also get information about how easily the child mixes in school with other children. But the hormone, once again, was showing relationships with social behavior at school age.

We were also interested in that concept of systemizing, how strong a child's interest was in systems of different kinds. Was this a child who liked to collect things, to have the complete set, for example, that makes up a system? Was this a child who loved to take things apart and put them back together again? So, very much interested in construction and assembly and figuring out how things work? Was this a child who spotted the small differences between different makes of cars, or little toy cars, and could tell you the differences between different varieties of system? Again, what we found (but this time the correlation was flipped over), was a positive correlation with prenatal testosterone. The higher the child's prenatal testosterone, the more interested they were in systems of one kind or another.

When they were about eight years old, we figured it was time to invite these kids into the MRI scanner, so that we could look directly at the question of whether testosterone is actually changing the way the brain is developing. Up until now we'd only done what's called behavioral studies, where we could find relationships between testosterone and behavior. But by eight years old a child is old enough to stay still, which is essential in a brain scan, because if the child squirms and moves around too much, then you can't interpret the results. These children, by eight, were able to tolerate having an MRI scan, and we were able to look at the structure of the brain and see if it had any relationship to prenatal testosterone. And in fact, there are lots of interesting relationships.

As I mentioned, one region of the brain that differs between males and females is that region called the planum temporale, a language area, and that's related to prenatal testosterone. The hormone is having an effect on the way the brain is growing, just looking at the volume of different regions in the brain. We looked at one other region, which is the corpus callosum. Some of you will have heard of this: it's the connective tissue between the two hemispheres, and the hormone was correlated with the asymmetry of one part of the corpus callosum, just toward the back of that structure. This was the first evidence that the hormone in humans is having an effect on brain structure, brain development, and it was mirroring what the scientists who do animal research had been telling us all along, but we needed to demonstrate it in humans.

I want to bring this back to my interest in autism: autism is a neurological disability, and it affects boys much more than girls, and we were interested in the possibility that that same hormone, prenatal testosterone, might be a risk factor, or might be playing some role in the outcome of autism. I told you that the sample of children we were studying was about 500 kids. And autism is a relatively rare condition. It affects about one percent of the population. What that means is within 500 kids, there might be five children who have autism, which would be way too small a number to be able to draw any inferences about prenatal testosterone and association with this medical diagnosis.

We've been tackling the problem in two different ways. The first way is that within these 500 children we've used a dimensional measure of autism, rather than looking at whether the child has autism or not. We've been looking at the dimension of autistic traits, where you can look at the number of autistic traits a child has. The assumption is that we all have some autistic traits and that autism isn't categorically different to the rest of the population. It's simply at an extreme of a bell curve, a normal distribution of autistic traits. There are ways to measure autistic traits. Most of them are questionnaires. You get the parents to fill out a questionnaire, which is basically a way of describing how many autistic traits their child has, and then independently, you can look at their prenatal testosterone and see if there's any correlation.

What we found—maybe some of you have anticipated the result of this particular stage of the research—is that the higher the child's prenatal testosterone, the more autistic traits they were showing at different stages in development, at two years old, and we've repeated that at four years old. That's telling us about prenatal testosterone and autistic traits. But what we'd really like to do, and this is bringing you right up to date with where our research is, is to see whether there's any relationship between prenatal testosterone and actually developing autism.

As I've already explained, having a sample of 500 kids doesn't allow us to answer that question, so we've been collaborating with scientists in Copenhagen, Denmark, at the Biobank there, because in Denmark they've been collecting amniotic fluid from women who have amniocentesis there. But they've been collecting those samples since 1980, so they have 100,000 samples sitting in their deep freeze. And for a scientist, that's a wonderful opportunity. The other thing that Denmark has is a national register for psychiatric diagnosis. Every time anyone in Denmark develops any kind of psychiatric condition, they're put into a central database, and so we know across the whole of Denmark who has autism. We can go to the Biobank and fish out their amniotic fluid if it's in there, to analyze it for prenatal testosterone.

Whereas in our normative study of 500 children, we were dependent on doing the first measure of the hormones in pregnancy and then waiting for life to unfold over a ten-year period, in this new study, we can work backward. We can say we know who's got autism, and we can look to see whether their amniotic fluid is still in the deep freeze. That then gives us, if you like, a fossil record of what their hormones were like, what their levels of testosterone were like when they were in the womb. It's reverse engineering the problem. And to just tell you where we're up to, we've just completed that study. The analysis is going on as we speak, so we hope to publish the results of that study later this

Simon Baron-Cohen

year. Effectively, that will tell us whether testosterone levels are elevated prenatally in children who go on to receive a diagnosis of autism.

This is a hormone that has fascinated me. It's a small molecule that seems to be doing remarkable things. The variation we see in this hormone comes from a number of different sources. One of those sources is genes. Many different genes can influence how much testosterone each of us produces, and I just wanted to share with you my fascination with this hormone, because it's helping us take the science of sex differences one step further, to try to understand not *whether* there are sex differences, but *what* are the roots of those sex differences. Where are they springing from? And along the way we're also hoping that this is going to teach us something about those neurodevelopmental conditions like autism, like delayed language development, which seem to disproportionately affect boys more than girls, and potentially have us understand the causes of those conditions.

Q & A

TOM STANDAGE: What's the thing that determines the level of the testosterone in the womb? What's the next step along the path?

BARON-COHEN: I mentioned that there are at least 25 different genes that can influence how much testosterone any of us produces. If we had been in that study, if we were part of that cohort, it's our genetic makeup in part that influences how much testosterone each of us produces. And these are genes that are common genes, they're not rare mutations, they're common genes we all have, but it just depends on which version of the gene you have.

They're polymorphic genes, different versions of the same gene. So that's one possible source of variation in testosterone levels, but there may be other sources too.

GEOFFREY CARR: How much of the variance between the sexes is captured by the variants in the testosterone levels? Obviously a male fetus is going to have more than a female. Is that a complete explanation of the difference?

BARON-COHEN: I'm very glad you've asked that. You're asking: is testosterone the only factor of interest? The answer is definitely not. It's one factor of interest. It may be accounting for about 30 or 40 percent of the variance, so it's not the only player in producing these differences. We could imagine that another set of very important influences would be genes on the X chromosome, so X-linked genes, just because we know that females have two X chromosomes, and us poor males only have one, and that may well make a big difference. So that would be a separate nonhormonal set of influences which . . .

GEOFFREY CARR: We all have X chromosome genes . . .

BARON-COHEN: We do, but there are various mechanisms which mean that if you have two copies of the X chromosome, the genes may work differently. One of them, for example, is called imprinting, which is a well-known mechanism where if you've got, for example, two X chromosomes, as females in this room are assumed to have, genes on one X chromosome can affect the homologous gene on the other X chromosome. That's an imprinting process, and that could change expression levels of proteins, for example. That would just be one example of a nonhormonal, non-

Simon Baron-Cohen

testosterone-related influence on brain development and behavior in humans.

PHILIP CAMPBELL: Are the animal models telling us about what the testosterone is actually doing, that's in the brain development?

BARON-COHEN: Yes. I didn't mention that, but I'm really glad you've raised it. Testosterone by itself doesn't really do anything, but for testosterone to do its work it has to bind to androgen receptors. There are these receptors that are just waiting for testosterone, and when testosterone is bound to the receptors, then it can start influencing all sorts of other processes. Androgen receptors are all over the body, but they're also in the brain and in some of those regions like the amygdala or the planum temporale. These regions are sexually dimorphic: they're different in one sex or another, and they're very rich in these androgen receptors. We know that from animal work, and people are beginning to create maps of androgen receptors across the brain. It's all about the hormone binding to these receptors.

Once it's bound, the hormone can do all sorts of things. It can, for example, modulate neurotransmitters. There's one called serotonin. There's another one called GABA. And testosterone seems to modulate how those neurotransmitters work. Testosterone also seems to affect connections that neurons make with each other. There's a process called apoptosis that some of you will have heard of, which is selective cell death, where neurons, or nerve cells in the brain, are pruned, so that we lose certain connections. Testosterone seems to affect the rate at which we lose those connections.

STANLEY BUCHTHAL: Are various regions factored in, because in the United States autism is rising rapidly?

BARON-COHEN: Autism is rising rapidly all over the globe. It's not just the United States. It's happening here, too. When I started out in the field of autism research, about 25 years ago, the reported rates of autism was four in 10,000 children. And everyone agreed that that was the rate of autism, about 1 in 2,500 kids. Whereas the reported rates of autism today are about one percent. It's gone from being very rare to being relatively common. Everyone's wondering what those increases in prevalence are due to.

Is it due to something in the environment? That would be one possibility. Or is it just due to better recognition and the fact that we are now trained to look for it, so we've got better diagnoses? My own view is it's probably explained by the latter: that there's been raised awareness of autism, and also there are more services on the ground for autism. There are clinics in almost every small town that can diagnose a child with autism, whereas 25 years ago there were probably only two or three centers in the country that were expert in diagnosing autism in this country. Just those ordinary factors, of more services, more awareness, and also some changes in how we classify autism. We've broadened the category so that we don't just pick up the severe forms of autism, but also milder forms. Those are the sorts of factors that I think, by themselves, could explain why we've gone from a relatively rare condition to a relatively common condition.

The reason I'm emphasizing these kind of mundane explanations for the increase in autism is that I'm sure we're all aware of the MMR theory, the measles, mumps, and rubella theory: the idea

Simon Baron-Cohen

that the vaccination that's used for MMR was in some way causing autism. That would have been an example of an environmental factor that was introduced at a certain point in time, and which
people thought was linked to the rise in autism. That theory has
been discredited, but not without doing a lot of harm. It raised a lot
of parental anxiety about whether to vaccinate their children, and it
actually had some other consequences that were really unfortunate:
that people didn't vaccinate their children and as a result the kids
ended up getting, for example, measles, which can be a very serious illness. So it had an impact on public health. So I think before
anyone puts forward an environmental factor that might explain the
increase in autism, I think what we learned from the MMR story
was that we should be very careful and only talk about potential risk
factors in the environment if there's very good evidence.

STANLEY BUCHTHAL: Is there any prevention in the near
future, or any cure?

BARON-COHEN: Prevention of autism, or cure of autism. Well,
the autism community, which includes the parents, those with the
diagnosis, the researchers, and the professionals, are very divided
on whether we should be trying to cure it, and whether we should
be trying to prevent it. It's a really interesting issue because, if
we were talking about some other kind of medical condition like
cancer, we would all be agreed that we should prevent it and we
should cure it and we should eradicate it. There wouldn't be any
kind of disagreement in the community because we know cancer
is a horrible disease. But when it comes to autism, autism is clearly
a disability—these are kids and ultimately adults who struggle to
socialize, struggle to make friends, struggle to keep friends, and
they face all kinds of challenges.

But that doesn't sum them up completely. They also have other characteristics, and many of those are not disabilities. In fact, sometimes they can even be talents. You can have a child who's very gifted at mathematics, or very gifted at building things, with a very strong interest in systems, and that same child who may be challenged when it comes to socializing may be doing wonderful, very novel, and very valuable things in other parts of their life.

So the idea that we would want to prevent autism or cure it? Autism is a complex medical condition. Certainly, there are symptoms of autism that we might like to treat, and I would, for example, focus on epilepsy, which develops in some people with autism, as an uncontroversial symptom that we would love to be able to eradicate or prevent or treat. Whereas the autism itself may be part and parcel of this child's nature, and some of that child's nature doesn't need treatment, doesn't need prevention—it actually needs to be fostered and valued and allowed to blossom.

MELANIE HALL: How confident can you be that the levels of testosterone that you find in the amniotic fluid are reflective of the levels of testosterone in the fetus? So that for example, the levels of testosterone in the fetus could be reflective of the level of testosterone in the mother in whose womb the amniotic fluid is. I'm interested in that relationship.

BARON-COHEN: Excellent question. The challenge we had as scientists is how to measure testosterone prenatally. Maybe I didn't make that clear enough. But if I wanted to measure testosterone in all of us right now, I could just get you to spit into a test tube and we could take your saliva off to the lab and analyze it for your *current* levels of testosterone. But all the theories are

Simon Baron-Cohen

telling us that it's not about your current level of testosterone, it's your *prenatal* testosterone that might be very important in understanding this variability in patterns of development. We somehow had to get at the child at that point in development, in an ethically acceptable way. We can't go into the fetus because that could do damage and there's that precept: "do no harm."

So these are women having amniocentesis anyway. I should also emphasize that it would be unethical to ask women to have an amniocentesis purely for research, because some of you may know that during amniocentesis itself, when the needle goes into the womb, it carries a small risk of inducing miscarriage, so the mother loses the pregnancy. It's fortunately a very small risk, but the risk is there. Effectively, what we recognized was that this was about as close as we could get to the fetus in an ethical way, to look at the hormones that were, in all likelihood, coming from the fetus.

The fetus is producing his or her own testosterone, and it gets excreted into the amniotic fluid. The mother, as you pointed out, is also producing her own testosterone, and the understanding is that testosterone doesn't cross the placenta, that there are enzymes in the placenta, one called aromatase, which converts testosterone to estrogen. So if you're finding any testosterone in the amniotic fluid, it's probably not come from the mother, and it's come from the fetus.

That doesn't completely answer your question, because you're also asking, "Do those levels reflect what's in the fetus's brain?" for example. But we just can't go there. We don't want to do anything that could risk the pregnancy. So it's the closest we can get, but we have to make an assumption that the testosterone you find in the amniotic fluid is a reflection of the testosterone levels in the fetus itself.

MELANIE HALL: Then how do you link that to what you said earlier about the need for receptors and relationships with testosterone to have the sort of impacts you've been describing? I suppose the question is "Are there any receptors in the amniotic fluid?"

BARON-COHEN: I'm sure that the room is divided on the topic of animal research. But for people who do find that it's acceptable to do that kind of research, those researchers are in a position where they can look at the relationship between testosterone in the amniotic fluid and testosterone in the animal itself. And actually, it does correlate pretty well. And they can also look at the receptors in, for example, the animal's brain, and look at the number of receptors at different points in development. Human research can't go as far or as quickly as what's possible in research with rats, for example. And I accept that even what I'm saying about research in rats would not be acceptable to some people. [Additional comment from SBC: Although there are no receptors for testosterone in the amniotic fluid, the testosterone is taken up by the blood, and there are receptors for it in many body tissues, including in the brain.]

ARMAND LEROI: Have you looked at people who might have had truly extreme exposure to testosterone in the fetus, or a complete absence of it? In other words, adrenal hyperplasia? I remember reading this book, a case study of a little girl who was born of a pregnancy in which in the third trimester the mother began growing a beard. And her daughter, who was born with strongly masculinized genitalia as a consequence of the vast overproduction of testosterone that turned out to be—the child turned out to be homozygous deficient for aromatase mutation, and all the testosterone that was produced by her placenta was not converted into estrogen, but just kept sort of . . . the placenta just kept pumping out testosterone, which is also what happens in adrenal hyperplasia. So there have got to be more people like that, or conversely, people that have androgen receptor deficiencies.

BARON-COHEN: I can tell you about our attempts to find them: that there are some medical conditions that could be very informative in terms of the role of testosterone prenatally. One of them is congenital adrenal hyperplasia. These are individuals who are overproducing testosterone for genetic reasons. Girls with CAH, as it's called, are of particular interest because it means that they've got two X chromosomes, just like most of the women in this room, we assume, but they're also producing more testosterone than the average female fetus. And we've looked at girls with that syndrome and we've given them tests of autistic traits. They do, indeed, have higher levels of autistic traits compared to their sisters who don't have CAH. In that kind of study, you're controlling for some background variables because they're in the same family, they share many of the same genes, but the main difference is that there was one sister with high levels of testosterone, and another sister in the same family without [high levels of testosterone]. That's giving us some clue that testosterone is influencing the number of autistic traits that a person has.

The opposite kind of medical condition is complete androgen insensitivity syndrome (CAIS), so these are chromosomally male fetuses. They have an X and a Y chromosome. They produce testosterone just like anyone with an X and a Y chromosome, but they lack androgen receptors. So we were talking earlier about the receptors. They produce the testosterone just like all of us to varying levels, but the testosterone can't do anything. It can't bind to the receptor. These are individuals who, despite having a Y chromosome, physically they look female, the parents assume that they are female, they're given girls' names, they're raised as girls, and everybody, including the child, believes that they are a girl until, typically, they get to puberty and at puberty they don't start to menstruate and they're taken off to the doctor for various tests. And then comes the momentous discovery that they have a Y chromosome, that they're actually chromosomally or genetically male, even though they look female in terms of the outward characteristics of the body. And these individuals would be very relevant to study because they haven't been able to use testosterone prenatally. You might expect that the patterns of interests that they have, the way their mind is developed, but also the way their brain is developed, is more like a typical female despite having a Y chromosome. And that kind of experiment, as far as I know, hasn't been done. I call it an experiment, but it would be research that's quasi-experimental. It would be looking at a group that arises naturally in nature to see what lessons we can learn for biology.

BRIAN ENO: In one of your books you characterized some autism as "extreme male brain" syndrome. Is there such a thing as "extreme female brain" syndrome? And what does it look like?

BARON-COHEN: Just to backtrack, the "extreme male brain" is kind of shorthand for someone who might have below average empathy, but a very strong interest in systems. If you imagine a child who really isn't interested in playing with any other child,

Simon Baron-Cohen

but loves to play with their LEGO set, because it's very systematic and they can build all kinds of wonderful constructions out of LEGO, but that when other kids try to interact and play with that child, the child's just not interested—that would be a kind of extreme of the male brain: totally focused on systems, disinterested in people.

You're asking about the converse: someone who might have excellent empathy, that their total focus is other people, but when it comes to understanding systems they're quite challenged. And that profile may well exist, but whether it would come to the attention of clinicians is not clear. What we know is that to the extent that autism fits that extreme male profile, we know that comes to the attention of clinicians, because by the time that child starts school they can't mix with other kids, they may withdraw from other kids, they may withdraw from teachers because they find people too difficult, and they may withdraw into a world of mathematics, which they find predictable and systematic, or physics, or fields which don't involve people but just involve patterns.

But the people with the opposite profile that you're describing, they might be excellent at socializing, so they wouldn't necessarily come to the attention of a clinician. They might find it difficult to do mathematics or to do subjects that involve systems, but you might just steer them toward other subjects in school.

BRIAN ENO: I ask the question because I knew a little girl who was so empathetic that she was almost paralyzed by her own empathy, because she was always calculating the effects of everything she did on everybody else. And she almost couldn't conceive of herself as separate from all of these other ramifications that her behavior would have, so it seemed to me like a disability actually. She had no individuality that she could deal with.

BARON-COHEN: What would be needed, speaking as a scientist now, is a whole group of people like this one girl. So what you'd need to do is identify people with a particular profile who seem to have excellent empathy on whatever measure you use. It might be questionnaires by their parents, it might be tests of how well they can read people's facial expressions, or tune in to whether somebody else is suffering or not. Excellent empathy, but who are simultaneously challenged by understanding systematic kinds of information.

What we'd want to know is whether that is a form of disability in its own right. What you're suggesting is it might, for some people, also be a disability because if you're constantly empathizing you might become over-involved in other people's feelings, maybe to your own detriment. That's a possibility, but that's why we would need to do the research. The other possibility is that maybe empathy is more like: you can never have enough of it. The more you have, the better your social networks, and that perhaps itself has a protective function, that you have more friends and you perhaps feel less isolated, less prone to depression through loneliness. We just don't know, but I think it would be a very good group to study.

MARY EVERS: The study group you used are women who had amniocentesis. So how relevant is that, because they tend to be a specific group of women?

BARON-COHEN: Absolutely right. This is a bit like the question we had earlier about "Is amniocentesis really the best opportunity or the best environment, I guess, to study these questions?" As you point out, women who have amniocentesis are not typical. In fact, it's only about 6 percent of women during pregnancy who choose to have an amnio. And usually it's because they're older than average. In this country, if you're pregnant and you're over the age of 35, your doctor will say, "I advise you to have an amnio," because maternal age itself is a risk factor for the baby having Down syndrome. But there are other reasons why your doctor might suggest you have an amniocentesis. We're already looking, from all the women who are pregnant, at just this small subset, the 6 percent who choose to have an amniocentesis.

It's not an ideal population to study. They're not a random population or a representative population. They're a special population. But they're the only ethical way that we can study hormones at the right point in development, to see what's going on in humans. If we were just to take a random sample of women who are pregnant, and ask them to have an amniocentesis, that would be unethical because we, as researchers, might be causing harm. We have to work within ethical limits and, thankfully, that's what science and human science does, it works within ethical regulated frameworks, and we have to try and think imaginatively as scientists, how to answer questions in an ethical way.

MARY EVERS: Is autism related to older parentage?

BARON-COHEN: There is some research coming out that is showing a link between parental age and autism. Some studies have found it's to do with the mother's age, older mothers being more likely to have a child with autism, and some studies are linking it to the father's age. It's too early to actually understand what age is reflecting.

But I should just go back to say that maternal age is something that we can try to control for in our studies. When we look at the child's hormone levels, we can also take out that part of the variance in the study that is to do with maternal age, because there is a spread of ages among the women that we study. We try to take that into account, but I recognized what you're pointing out is a limitation of this kind of research.

MARCUS DU SAUTOY: I have two genetically identical girls, but one is a real systemizer, loves doing LEGO; the other one's big on empathy. Do you have any interesting examples of twin studies where you see, where the testosterone level is the same that they've been exposed to, but the results seem very different?

BARON-COHEN: There are a couple of things to say about that. The first thing is that in our study of 500 kids we excluded twins, and we did that for a reason, which is that when amniocentesis takes place and they put the needle in, it's never very clear which amniotic sac that surrounds the fetus the fluid has come from. Was it twin A or twin B? Because there might have been some ambiguity, some doubt about whether this sample of amniotic fluid belonged to twin A or twin B, we thought: let's exclude twins. And also, because, although some twins are in their own amniotic sac, some twins actually share the same amniotic sac. So there's a lot of extra complexity that comes into that. So our study just looked at what are called singletons: as in, there was only one baby in that pregnancy.

But the other thing you're pointing out is that you've got these beautiful twin daughters who are very different, despite the fact that they are, to all intents and purposes, genetically identical.

That's telling us that even when you have two genetically identical people, your daughters, they end up developing differently, which means that genes don't determine our behavior 100 percent.

And we can say that just from your single case study, your twins . . . but we can also say that on the basis of a national study of twins that goes on in this country and in many countries, where they've taken large sets of twins and looked to see how much of human behavior can be explained by genes. And actually whether you're looking at empathy, or systemizing, or many other aspects of human behavior, genes really only account for about 50 percent of the variance. So that's to say that environmental factors are also important, and gene-environment interactions. It's more complicated than simply "genes explain our behavior," and I want to underline the importance of nongenetic environmental factors in shaping our interests.

MARCUS DU SAUTOY: There is this general feeling why mathematics is a male-dominated subject. Certainly, going to mathematics conferences, it's very male. My feeling is that people try to put it on biology, your interesting study about the 24-hour children, that they were already responding to that. I still think it's quite difficult . . . your answer just now illustrates that the environmental considerations might mean that that's why mathematics departments are still so male-dominated. But that's still a much stronger effect than the structure of the brain.

BARON-COHEN: We're really only at the beginning of this kind of research. I described a study that we've been doing for 10 years. Amniocentesis has only been around since the 1980s and the animal work goes back to probably about the 1940s. But I would say that we're still only at the beginning of understanding this area. And the issue about what the hormone is doing, and how that interacts with genes: there's still a lot to understand there. So testosterone, it's now known, can also affect whether genes are turned on or turned off. It's got this epigenetic characteristic too. It's not just that you have to look at genetic influences separate from hormonal influences, but whether the hormone can actually influence the timing of whether genes are turned on and turned off. We still have a lot of work to figure out.

But you mentioned mathematicians, and I think you're right, that there are these areas of human activity—math is one of them—where we do see very disproportionate sex ratios. My understanding is that in mathematics, at university level, it's about 14 males for every one female sitting in the audience in those lectures. That's a very big difference. And there are other sciences, as we know, that used to be like that but that have changed dramatically. Medicine is a very good example. It used to be male-dominated and it's now certainly 50–50, or if anything, it's gone beyond and there are now more female applicants and thankfully, successful applicants. If you look at the audience in medical lectures, the sexes are, if not equally represented, maybe even more women than men. But there remains this puzzle why mathematics, physics, computer science, engineering, the so-called STEM subjects, why they still remain very male biased.

I'm the first to be open to anything we can do to change the selection processes at university, or change the way we teach science and technology at school level, high school level, to make it more friendly to females, to encourage more women to go into these fields. But there remains a puzzle as to why some sciences are attracting women at very healthy levels, and other sciences, including mathematics, remain much more biased toward males. Whether that's reflecting *more* than just environmental factors, and something about our biology, is something that I think we need to investigate.

DAVID ROWAN: If your correlation is confirmed, I can see an opportunistic pharmaceutical company targeting anxious pregnant women with a product that claims to help you modulate your child's hormones, maybe extra aromatase. Do you see an ethical conundrum here?

BARON-COHEN: Yes. The field I'm working in, prenatal hormones, is plagued with ethical conundrums. And while we are doing the research, I've got to keep one eye beyond the results. The results are not yet available and so the research is not yet complete. But I have to be thinking one step ahead: that when the research is available and complete, what are the real-world ways in which that research could be used by a pharmaceutical company, or by others? And some of those ways might include, as you've said, that a company might say, "We can now test if your baby is going to have autism." That might be one scenario if the test was highly specific and strongly predictive. And those are big ifs. But if such a test existed, you could imagine that there would be interest by companies to market such a test.

Another ethical minefield is treatment: that if it turns out that hormone levels, prenatally, are involved in shaping the way the mind and brain develops, and that too high a level can tip a child into autism (and this is still an if, because we're not yet at that point), you could easily imagine that some companies, pharmaceutical companies, or some doctors, would want to try to intervene: for example, by using testosterone blockers to manipulate the child's hormone levels prenatally, or at birth. And again, this frightens me enormously because we know that testosterone, if it is involved in autism (and that's still an if), is also involved in many other things. It's involved in bone development. It's involved in gender identity. It's involved, potentially, in other psychological processes, like spatial ability, maybe mathematics. Once you start playing with the hormones, what are the side effects? Are you just targeting risk of autism? And it relates to the question earlier about "Should we even be treating or trying to prevent autism?" But also, once you start playing with hormones, you might be playing with a number of different systems with unpredictable outcomes. I'd be quite worried.

But at the same time, the research is being done to just try to understand, taking one hormone at a time, to see how it plays out. We've got the activity of the scientists who are doing their laboratory experiments, or in this case, human studies, but we have to think ahead about ethical debates that might be needed if results turn out the way the studies are predicted.

BRUCE HOOD: If testosterone is a guilty culprit, and we know it's controlled by, say, 25 genes, then why has the genes analysis eluded us when it's come to understanding autism?

BARON-COHEN: This is another good question. Autism genetics, like the genetics of many psychiatric conditions, has tended to proceed along what's called genome-wide association studies

Simon Baron-Cohen

(GWAS) where you look at all the genes in the human genome and you set a level of probability . . . I'm talking to you as a fellow scientist . . . you set quite a "severe" level of probability of what's going to count as a significant association. Usually it's something like 10 to the minus 7. It's not the usual p value that most scientists chase of .05—it's 10 to the minus 7. So it's a very strict level of significance before you can say this gene is associated with condition X. And what's found is that some genes pop up when you set the bar that high, and the rest are excluded from the analysis. It may be that genes involved in testosterone just haven't met that bar, and it may be that there are gene-gene interactions: that one gene by itself is not going to cross that bar, but genes working in combinations might. And that kind of analysis is hardly ever carried out in psychiatric genetics, looking at interactions between genes.

ED HALL: I should consider some human relationships, particularly in adults, to be a kind of system, and if some people with autism who could perhaps analyze systems particularly well could slip through undetected?

BARON-COHEN: Yes. That's absolutely right, and that could work for certain aspects of human behavior. Let's take lawyers as an example. Lawyers work using a system, the legal system, and they apply their system to a lot of human behavior, like the landlord and the tenant, or the employer and the employee, and there are lots of laws that govern the way that these different people behave. Some systems can help make sense of and even regulate human behavior, and you could also imagine that if you were very good at learning systems, you might be able to learn how to negotiate some areas of human behavior. Let's take salsa dancing as an example, where there are very strict rules to how two individuals relate to each other, and who leads and who follows on the dance floor.

So there'll be areas of human behavior where systemizing could really work, the law, dance, the military, where human behavior is governed by rules. There will be other areas of human behavior, like having a conversation, which are very difficult to systemize. It's very difficult to predict how a conversation is going to go if you're just using a set of rules, and if you expect that there are going to be patterns. A systemizing approach to human behavior could take you quite far in certain areas of understanding people. But it might not be adequate, or up to the task, in understanding human behavior in *any* environment.

WILLIAM McEWAN: Talking about scale of this autistic spectrum with empathizing behavior coming at the expense of systemizing behavior, what's the sort of limitation on having both, and is there a limitation on having both, and why not?

BARON-COHEN: Well, empathy is a dimension and systemizing is another dimension, and we do all have both—it's just about how much we have of each. And I think you're asking: "What about individuals who have a high level of both? Is there any downside to that, or is that actually the optimal kind of profile to have?" This relates to Brian's earlier point, of having different profiles. He was thinking about someone who had very good empathy, but maybe very poor systemizing. You're now picking out someone whose empathy and systemizing might be at high levels, where both of them are high.

Simon Baron-Cohen

I can't see any potential downside to a profile like that. It means that this is someone who finds socializing very easy, reading people is very easy, but equally, if they had a math textbook, they could make sense of it, or if they need to lift the bonnet of their car they could start figuring out systematically what one component might be involved in. Having a good level of both I could see might be a sort of optimal state, but those people are in the minority. What we tend to find is that whilst there are people who are equally good at systemizing and at empathy, most people show a bias one way or the other, and it's very interesting as to why populations ended up with that kind of distribution.

PHILIP CAMPBELL: I think the diagnosis of autism is purely behavioral, correct me if I'm wrong.

BARON-COHEN: That's correct.

PHILIP CAMPBELL: When do you think this is getting into the ethical question as well—when do you think biological contributions are going to come to diagnosis? When is that going to fundamentally change the ethical dilemmas? We all know that's crystal ball stuff.

BARON-COHEN: Right. In the majority of psychiatry, the diagnosis rests on behavior. So if you want to determine that someone has schizophrenia or if someone has depression, you're asking people about their own symptoms, so it's self-report (or parent-report), and you're looking at their behavior. And autism is no different.

At risk of saying something controversial, perhaps the limitation of psychiatry as a discipline is that diagnosis rests on behavior, and there are many different ways that you can interpret behavior. Many of us hope that one day diagnosis in psychiatry will be more like diagnosis in the rest of medicine, where you run a bunch of tests on biology, on "biomarkers," and you get a more objective output from those tests.

And the question is whether that will ever actually come to be. Some people would say: even if we knew all the genes and all the hormones that contribute to, say, autism, maybe having that particular set of markers still doesn't mean that you need a diagnosis. We might be carriers of those genes, for example, but we find ourselves in certain environments where those genes are not interfering with our functioning; and there will never be a day when you could substitute a purely biological test for a clinical interview, having a chat with a doctor where they say, "And how are you?" And if you say, "I'm having a hard time," then you might meet the diagnosis. And if you say, "I'm fine," despite having this genotype, why need the diagnosis?

It may be that biomarkers will improve and maybe they'll be an adjunct to diagnosis. That might be the best possible outcome, that they *assist* us in making diagnosis more objective. But at the same time, we might still need that human contact, to just check "Does this person actually need a diagnosis, in the sense of need help?" Because that's really what the diagnosis is all about.

SEIRIAN SUMNER: I was just wondering whether you could comment on the plasticity of the two different ends of the spectrum, empathy versus systemizing. You can't learn empathy—maybe you can try and improve your empathy skills but it's really innate, you can't really be taught it—whereas you can be taught how to systemize systems, you can be taught how to do mathematics, you can be taught how to categorize, or lift up the bonnet of your car.

BARON-COHEN: Sure. I might slightly disagree with you, politely, in the sense that there's probably plasticity in many aspects of human learning and empathy might be . . . it's not that it's no different from other kinds of learning, but there still is a learning element. You were saying "empathy is innate," and so that means that you're born with as much empathy as you're ever going to have, that the genes just have to switch on and that would determine how much empathy you have. Whereas I think probably what a lot of our research from child development has taught us is that there are major environmental influences to empathy, and I would just point to the work of John Bowlby.

Many of you will have heard of him, a child psychologist who studied medicine at Trinity College, my college in Cambridge, and who went on to develop "attachment" theory, which looks at the attachment between the baby and the parent, usually the mother. He basically found that the quality of that relationship between mother and baby or father and baby predicts quite a lot about how much empathy the child ends up with. Children who are abused or children who are neglected, or children who are just anxious about their relationship with their parent, with "insecure attachment," tend to also find it much more difficult to trust other people and to read other people, to develop empathy, in later life. That would just be one example of plasticity, as you called it, in the development of empathy.

We were talking earlier about the twins, how much can innate factors explain about human behavior. I keep in mind, as a rule of thumb, about 50 percent. The other 50 percent is about experience, is about upbringing, and these factors interact: genes, hormones, and our experience all interact, and when we try to be reductionist, or try to be too simplistic, to emphasize the role of one of these factors, actually we're missing the complexity of how they interact.

10
Insight

Gary Klein

Senior Scientist at MacroCognition LLC; he was instrumental
in founding the field of naturalistic decision making; Fellow,
American Psychological Association and the Human Factors and
Ergonomics Society; author, *The Power of Intuition* and *Seeing
What Others Don't*.

INTRODUCTION by Daniel Kahneman
Recipient of the 2002 Nobel Prize in Economics; Eugene Higgins
Professor of Psychology Emeritus, Princeton University; author,
Thinking, Fast and Slow.

*We are prone to think of the British as snobbish, a label that is rarely
used to describe Americans. When it comes to the adjective "applied,"
however, the tables are turned. The word "applied" does not have any
pejorative or diminishing connotation in Britain. Indeed, the Applied
Psychology Unit on 15 Chaucer Road in Cambridge was for decades the
leading source of new knowledge and new ideas in cognitive psychology.
The members of that unit did not see their applied work as a tax they had
to pay to fund their true research. Their interests in the real world and in
theory merged seamlessly, and the approach was enormously productive
of contributions to both theory and practical applications.*

*In the United States, the word "applied" tends to diminish anything
academic it touches. Add the word to the name of any academic discipline, from mathematics and statistics to psychology, and you find lowered status. The attitude changed briefly during World War II, when*

the best academic psychologists rolled up their sleeves to contribute to the war effort. I believe it was not an accident that the 15 years following the war were among the most productive in the history of the discipline. Old methods were discarded, old methodological taboos were dropped, and common sense prevailed over stale theoretical disputes. However, the word "applied" did not retain its positive aura for very long. It is a pity.

Gary Klein is a living example of how useful applied psychology can be when it is done well. Klein is first and mainly a keen observer. He looks at people who are good at their jobs as they exercise their skills, sometimes in life-and-death situations, and he reports what he sees in clear and eloquent prose. When you read his descriptions of real experts at work, you feel that it is the job of theorists to accommodate what he has seen—instead of doing what we often do, which is to scan the "real world" (when we think of it at all) for illustrations of our theoretical notions. Many of us in cognitive and social psychology are engaged in the important exercise that Lee Ross has wonderfully described as "bottling phenomena," and our theories are built to fit what we bottle. Klein himself is a theorist as well as an observer, but his theoretical ideas start from what he does for a living: they are intended to describe and explain a large chunk of behavior in a context that matters. It is instructive to note which of the concepts that are current in academic psychology turn out to be useful to him.

Klein and I disagree on many things. In particular, I believe he is somewhat biased in favor of raw intuition, and he dislikes the very word "bias." But I am convinced that there should be more psychologists like him, and that the art and science of observing behavior should have a larger place in our thinking and in our curricula than it does at present.

What's the tradeoff between people using their experience (people using the knowledge they've gained, and the expertise that they've developed), versus being able to just follow steps and procedures?

We know from the literature that people sometimes make mistakes. A lot of organizations are worried about mistakes, and try to cut down on errors by introducing checklists introducing procedures, and those are extremely valuable. I don't want to fly in an airplane with pilots who have forgotten their checklists and don't have any ways of going through formal procedures for getting the planes started and handling malfunctions, standard malfunctions. Those procedures are extremely valuable, and I don't doubt any of that. The issue is, how does that blend in with expertise? How do people make the tradeoffs when they start to become experts? And does it have to be one or the other? Do people either have to just follow procedures, or do they have to abandon all procedures and use their knowledge and their intuition? I'm asking whether it has to be a duality. I'm hoping that it doesn't, and this gets us into the work on system-one and system-two thinking.

System one is really about intuition, people using the expertise and the experience they've gained. System two is a way of monitoring things, and we need both of those, and we need to blend them, and so it bothers me to see controversies about which is the right one, or are people fundamentally irrational and therefore they can't be trusted. Obviously, system one is marvelous. Danny Kahneman has put it this way: "System one is marvelous, intuition is marvelous but flawed." And system two isn't the replacement for our intuition and for our experience; it's a way of making sure we don't get ourselves in trouble.

If we eliminate system one, system two isn't going to get the job done because you can't live by system two. There are people who try—there are people who have had various kinds of brain lesions that create disconnects between their emotions and their decision-making process. Antonio Damasio has written about them. It can take them 30 minutes to figure out what restaurant they want to go to. Their performance on intelligence tests isn't impaired, but their performance in living their lives is greatly impaired; they can't function well, and their lives go downhill.

So we know that trying to do everything purely rationally, just following Bayesian statistics or anything like that, isn't going to work. We need both system one and system two, and so my question is, what are the effective ways of blending the two? What are the effective ways that allow people to develop expertise, and to use expertise while still being able to monitor their ideas, and monitor their actions?

Too often it's treated as a real dichotomy, and too many organizations that I study try to encourage people to just follow procedures, just follow the steps, and to be afraid to make any mistakes. The result is that they stamp out insights in their organization. They stamp out development of expertise in their organization, and they actually reduce the effectiveness and the performance of the organizations. So how you blend those is an issue.

A project that I'm working on now is about what to do with children who are in dangerous situations in home environments where a Child Protective Service worker has to judge the potential for abuse. Do you leave the child with the parents who do damage to the child, or do you remove the child for its own protection, which creates its own set of problems.

A complicating factor here is that a lot of the Child Protective Service workers are not all that well paid; there is a lot of turnover; they don't develop expertise, so there is a temptation to turn it all over to checklists and say, "Here are the factors, these are the things to go through, these are the objective criteria." Some of the criteria are useful and need to be taken into account, but there must also be an aspect of empathy, of a caseworker looking at the situation and saying, "This doesn't feel right to me. There is an

Gary Klein

edge to the way the parents are interacting with the child, there is a feeling of hostility I'm picking up, there is a feeling of menace in the way the parents are acting, and the way the child is acting," and I don't think we want to lose that. I'm afraid that the temptation to try to procedurize and checklist everything can get in the way of those kinds of insights, and those kinds of social concerns that seem so important.

I started moving in this direction when I began working as a research psychologist for the Air Force, at Wright-Patterson Air Force Base. I worked for an organization called the Air Force Human Resources Laboratory starting in 1974. I had been in academia, and I enjoyed that, but this was an opportunity to do something different. It was a wonderful time to be at Wright-Patterson and in my office because the Arab oil embargo had hit in 1973, and the price of fuel went sky-high, and jet fuel went sky-high, and all of a sudden these pilots who were used to doing all their training by just flying, because they loved to fly, that's what really got them so excited, now they were going to have to develop skills and be trained in simulators. My unit was studying simulation training. Up to that point pilots really didn't care much about what the unit did because they just wanted to fly. Now it mattered because they were going to spend a lot of time in simulators building their skills, and so this meant a lot to them.

For me, it was a chance to confront some basic issues about the nature of expertise, because we were trying to develop expertise in the pilots in an artificial setting, which is not as good as being able to fly. But in some ways it's better because you can practice certain maneuvers, you can confront them with certain malfunctions that you can't do in the air. It opened up the question, what is the nature of expertise, how does it develop, how can we use these kinds of devices to develop the pilots more effectively?

We did that for a couple of years, and as I was doing that, a part of the expertise that intrigued me was how people make tough judgments, and how they make decisions. I wanted to get involved in that, I wanted to study that. There wasn't any place for doing it in my branch, so I left and started my own company in 1978 to examine those sorts of issues, and struggled for a while to try to see where I was going to go with this to get funding.

Then in 1984, a notice came out from the Army Research Institute asking for proposals about how people make life-and-death decisions under extreme time pressure and uncertainty. We said, "That sounds like the sort of thing that we wanted to get involved with." We knew that the standard way you do that kind of research is you pick a laboratory task, a well-studied, well-understood task, you vary the time pressure, you vary the uncertainty, and you see the effect. But we didn't want to do that, because according to all the literature we had seen, it should be impossible to make good judgments and good decisions in less than a half hour. Yet we know people can do it and we didn't know how.

Rather than perform standard research and manipulate variables, we said, "Let's talk to the experts." We found some experts; we found firefighters. We decided we would study firefighters because that's what they do for a living, that's how they've trained themselves, that's how they develop themselves, and because we figured that when they weren't fighting fires, they would have time to talk to folks like us. There was another advantage that we didn't know at the time, which is that firefighters are a wonderful community to work with. They're very friendly, they're very open to being helpful. We went in and said, "The Army wants to do a better job of preparing people to make tough decisions. You guys are the experts—are you willing to share what you've learned?" And they said, "We'd love to help." They were marvelously cooperative.

That's the way we began the research. We got funded, and then we had to go out and see what the firefighters were doing. We started out in the wrong direction. At first, we thought we would just ride along with them. But they don't have enough interesting two-alarm, three-alarm fires, and we were going to waste all of our energy and all of our funds sitting around in fire stations, so we said, "Let's not do that, let's interview them about the tough cases that they've had."

I remember the first firefighter I interviewed—this was just a practice interview—and I said, "We're here to study how you make decisions, tough decisions."

He looked at me, and there was a certain look of not exactly contempt, but sort of condescension, I would say at least, and he said, "I've been a firefighter for 16 years now. I've been a captain, commander for 12 years, and all that time I can't think of a single decision I ever made."

"What?"

"I don't remember ever making a decision."

"How can that be? How do you know what to do?"

"It's just procedures, you just follow the procedures."

My heart sank, because we had just gotten the funding to do this study, and this guy is telling me they never make decisions. So right off the bat we were in big trouble. Before I finished with him, before I walked out, I asked him, "Can I see the procedure manuals?"

I figured maybe there's something in the procedure manuals that I could use to give me an idea of where to go next. He looked at me again with the same feeling of sort of condescension (obviously I didn't know that much about their work) and he said, "It's not written down. You just know."

"Ah, okay, that's interesting."

Something was going on here. It feels like procedures to them, but it's not really procedures, and it's not that they're following the steps of any guide, or any set of checklists. We conducted a few dozen interviews to what people were doing, and we collected some marvelous stories, and some really very moving stories from them about how they made life-and-death decisions. What we found was that they weren't making decisions in the classical sense that they generated a set of options, and they looked at the strengths and the weaknesses of each option, and then they compared each option to all the others on a standard set of dimensions. I mean, that's classical management-type decision making, get your options, A, B, and C, get your evaluation dimensions, rate each option on each dimension, see which comes out ahead. They weren't doing that.

When I asked him, "How do you make decisions?" that's what he assumed I was asking about, and they never did it, and that's why he gave me that response. Others gave me that response, and pretty quickly we stopped even asking them about how they made "decisions." Instead, we would ask them about "tough cases," cases where they had struggled, cases where they maybe would have done things differently, cases where they might have made mistakes earlier in their careers. We were asking it that way to get away from the term "decision making," because that was just leading us in the wrong direction.

I'll give you a few examples. One example, a firefighter insisted he never made any decisions and so I didn't even push it any further.

I said, "Tell me about the last run you went on."

And he said, "It was about a week ago."

"Tell me what happened."

I figured maybe we could learn something from it. He said, "It was a single-family house, and we pull up, and I see there is some smoke coming out from the back.

Gary Klein

"Right away," he says, "I figure it's probably a kitchen fire. That's a typical thing that goes on here. And I know how to handle kitchen fires."

He starts to walk around the house, but he's pretty sure he knows what's going on, so he sends one of his crew in with an inch-and-three-quarter line. He says to the men, "Go into the house, go in the back, and get ready, and when you reach the fire, just knock the fire down." That's what they did, and they put the fire out.

He said, "You see, there were no decisions, it was all just standard."

I said, "You know, I've always been told if a house is on fire, you go out of the house. And you just told me you sent your crew into the house. Why did you do that? Why didn't you just go around the back, break a window, and use your hoses from the back and hit the fire that way? That way none of your crew is inside a burning house. That seems to be the way I'd do it."

He looked at me with a little bit of condescension (I get a lot of condescension in this line of work) and he said, "Hm, yeah, maybe some volunteer fire companies might do it that way, but it's a bad idea. Let me tell you why. If you do it that way, what are you doing? You're pushing the fire back into the house, and now you're spreading it further into the house, and we don't want to do that. That's why we want to go into the house, get to the fire, and then drive it out. We want to hit it with water, and all the momentum, all the direction is pushing the fire out of the house, and not into the house. Now, there are times when we can't do that kind of an interior attack—like if there is another house right next to it that we could set on fire, then we would do it externally. But in this case, there weren't any complicating factors."

I said, "Thank you, thank you."

A simple situation. But, in fact, he had encountered a decision.

Do you do an interior attack, or an exterior attack? There was a decision point. He didn't experience it as a decision point because for him it was obvious what to do. That's what 20 years of experience buys you. You build up all these patterns, you quickly size up situations, and you know what to do, and that's why it doesn't feel like you're making any conscious decisions—because he's not setting up a matrix. But that doesn't mean that he's not making real decisions because the decision I would have made, he thought was a bad choice in this particular situation.

That became part of our model—the question of how people with experience build up a repertoire of patterns so that they can immediately identify, classify, and categorize situations, and have a rapid impulse about what to do. Not just what to do, but they're framing the situation, and their frame is telling them what the important cues are. That's why they're always looking, or usually looking, in the right place. They know what to ignore, and what they have to watch carefully.

It's telling them what to expect, and so that's why performance of experts is smoother than the performance of novices, because they're not just doing the current job, they know what to expect next, so they're getting ready for that. It's telling them what the relevant goals are so that they can choose accordingly.

Sometimes you want to put a fire out, and sometimes the fire has spread too much and you want to make sure it doesn't advance to other buildings nearby, or sometimes you need to do search and rescue. They've got to pick an appropriate goal. It's not just: put the fire out each time.

It's also telling them what to expect, and by the way, when they think about what to expect, that gives them another advantage, because if their expectancy is violated, that's an indication, "Maybe I sized it up wrong. Maybe my situation awareness is wrong, maybe the way I've made sense of it is leading me in a wrong direction, I framed it in the wrong way, and I've got to rethink it."

We saw many examples where they would be surprised because their expectancy would be violated, and that would stop them in their tracks. That's the importance of a frame; it gives you expectancies so that you can be surprised. The frame is also telling them what's the reasonable course of action.

But that's only part of the decision-making story, because that gives you your initial intuitive impulse about what to do. How do you evaluate that course of action?

When I started the research, I always assumed that the way you evaluate a course of action is that you compare it to other courses of action, like in the standard model, to see which one is better. But these firefighters were making decisions in just a few seconds, not enough time to evaluate by comparing to other options. Besides, they were telling us that they didn't ever compare it to other options. So what was going on there?

We looked at some of the cases, and we got an indication of what was happening.

I'll give you an example. A harness rescue example is one that I learned a lot from:

A firefighter gets called out, there's an emergency. It seems that a woman has decided to kill herself. She went to an overpass over a highway, way high over the highway, and she jumped off to try to kill herself, but she missed. Instead of falling to her death, which was her plan, she just fell a little bit, and she landed on the struts that were holding up one of the highway signs, and she reflexively grabbed onto the strut. The firefighters were called in, and the emergency rescue squad was called in, to try to save her. They pull up, and they see her there, and now the fire commander has to figure out, "How am I going to do this rescue?"

In the meantime, another piece of equipment, a hook and ladder truck, had been called in, and they radioed to him,

"How do you want us to help you?"

He didn't want to think about them, he just wanted to get them out of his hair, so he said, "Go drive down to the highway below, block the highway. In case the woman falls, we don't want her falling on top of a car, killing a motorist, you just block them."

Now he's got to figure out how to make the rescue. The standard way firefighters make a rescue is a Kingsley harness: you snap it onto the shoulders, you snap it onto the thighs, and you lift the person up to safety. You've seen it on TV—this is how they make rescues and get people out of dangerous situations. They lower the Kingsley harness, the person snaps it on, and they lift them up to safety. It wasn't going to work here because this woman was also either on drugs or alcohol, or both. She was semiconscious.

I forgot to tell you one other part of this story: as soon as he got there, the commander told his crew, "Nobody go out there, because it's too risky, it's too dangerous."

Two of his crew members immediately ignored him. They climbed out to try to protect her. One was sort of holding her legs, one was holding her arms. The commanders was glad that they were doing that because it was keeping her secure, and if anything happened to them, everybody had heard him tell people not to do it. So he was sort of covered either way. Now he's got two firefighters with her, and he thought, "Can we have them lift her up to a sitting position and then snap on the Kingsley harness?" But they're also balancing on those struts. There's not going to be any easy way for them to lift her up without making it really risky for all of them. So he rejects that option, that's not going to work.

Then he says, "You know, maybe we can attach the Kingsley harness from behind." She was facedown.

So he thinks, "We can try that."

But then he imagines what would happen as they lifted her, and he imagines the way her back would sag, and it was a painful vision, a painful image, and he thought, "Too much of a chance we're going to do damage to her back, that's a bad option."

He rules that out. He thinks of another few options, and when he imagines each of them, they're not going to work, so he rejects all of them.

Then he has a bright idea, he has a clever idea. What about a ladder belt? A ladder belt is what firefighters wear. Ever see on TV a firefighter on a second or third floor, and they've got a hose, or they're helping people onto a ladder? The ladder is up two or three stories high. They're at the top.

You think, "What if these guys fall?"

But what happens is they have a ladder belt that they've cinched up around their waist, and when they get to the top, there's a snap, and they snap it to the top rung, so they're well-secured.

The commander thinks, "We can just get a ladder belt. We can just lift her an inch, slide the ladder belt underneath her, buckle it from behind. We have a rope, we'll tie a rope to it, and we lift her up, and we can lift her to safety that way."

And he imagines it—he does what we call a mental simulation. He sort of works it through in his head to see if there will be any problems, and he can't think of any. So that's the way he makes a decision to do the rescue.

They lower the ladder belt; the firefighters slide it underneath her and tighten it up. In the meantime, the hook and ladder truck below is getting bored because they're just standing there, they're not doing anything. So they put somebody in the ladder, and they're raising it to do the rescue, and they're saying, "We're getting there, we're going to be able to rescue her."

So there's a race because he wants to get her rescued before they get there.

They cinch up the ladder belt, they tighten the buckle, and they start to lift, and that's when he realizes what was wrong with his mental simulation. Because a ladder belt is built to fit over a sturdy firefighter on the outside of their coat, which is a protective coat, and she was a slender woman. She was wearing a very skimpy sweater, and the firefighters tightened it to the last hole as tight as they could make it, and it wasn't tight enough. As they were lifting up, and I'll never forget his phrase, he said, "She slithered through like a slippery strand of spaghetti." She's sliding through. But the ladder people from below were right there, and as she's sliding through, they're rushing over and maneuvering the ladder that they were raising, and they catch her as she's sliding through, they make the rescue, and the woman is saved. So it worked, but not the way he expected.

What did we learn from that episode, that incident?

We learned about how he did the evaluation. He looked at several options, but he never compared them to see which was the best one compared to the others. He wanted the first one that would work, and he did this mental simulation. He did this imaging process for each one. That was the way he could evaluate one option at a time, not by comparing it to others, but by seeing if it would work in this particular context. Even though this didn't have the outcome he expected, that turned out to be another part of the decision-making strategy that the firefighters were using.

It's really a two-part strategy. The first part is the pattern matching to get the situation framed about what to do. Then the second part is this mental simulation to be able to evaluate and monitor an option to make sure that it will do a good job, and to use your experience. They would use their experience to do the mental simulation.

In this case, the firefighter, his experience didn't go far enough, and so his mental simulation didn't work as well as he would have liked. Generally, though, it's a very powerful technique.

What I've described about their strategy is about how they use their intuition, because they're not making formal decisions, they're not making analytical decisions by comparing options. These are intuitive decisions, and by intuition here, I'm talking about the way they are able to use their experience. This isn't just "top of my head, this feels good" type of decisions. These are intuitions that are based on 10, 15, 20 years or more of experience that has allowed them to build a repertoire of patterns that allows them to quickly frame situations, size situations up, and know what to do. But they're not just using intuition—they're balancing the intuition with the mental simulating part, which is how they're doing the analysis. So their decision making—we call it recognition-primed decisions. The decisions are primed by their ability to recognize situations, and balanced by the monitoring of the mental simulation.

The concept of intuition gives a lot of people pause because intuition feels magical, it feels like Luke Skywalker getting in touch with The Force, and we're not talking about that, about people somehow drawing on pyramid power, or something occult, or ESP. Although I've had a couple people I've interviewed tell me they made decisions because of ESP, because they couldn't articulate the basis of the decisions.

Intuition is about expertise and tacit knowledge. I'll contrast tacit knowledge with explicit knowledge. Explicit knowledge is knowledge of factual material. I can tell people facts, I can tell them over the phone, and they'll know things. I can say I was born in the Bronx, and now you know where I was born. That's an example of explicit knowledge, it's factual information.

But there are other forms of knowledge. There's knowledge about routines. Routines you can think of as a set of steps. But there's also tacit knowledge, and expertise about when to start each step, and when it's finished, when you're done and ready to start the next one, and whether the steps are working or not. So even for routines, some expertise is needed.

There are other aspects of tacit knowledge that are about intuition, like our ability to make perceptual discriminations, so as we get experience, we can see things that we couldn't see before.

For example, if you ever watch the Olympics and you watch a diving competition, the diver goes off the high board, and the TV commentators are there and the person didn't do a belly flop, dove in, looks clean, and they're saying, "Look at the splash. The splash was bigger than it should have been, the judges are sure to catch that," and what happened was the diver's ankles came apart just as she was entering the water. Then they show it in instant replay, and sure enough, that's what happened. But the commentator saw it as it happened. To a viewer like me, that's invisible. I just saw the dive. But they know where to look, and they know the probable trouble spots, or they know the difficult aspects. That's part of the patterns that they've built up—to know how to direct their attention so they can see the anomalies. They see it as it's happening, not in replay. You can't tell somebody over the phone what to look for. You can say it after the fact, but they see it while it's going on. That ability to make fine discriminations is a part of tacit knowledge, and a part of intuitive knowing.

Another part is pattern recognition. If you go to a friend's house, and the friend for some reason has an album out, and there's a picture from when they were in the fourth grade, you can look at the picture, and you look at all the faces, and you say, "That's you, isn't it?" And most of the time you get it right. Now, the face doesn't look like the face of your friend right now, but we see the facial re-semblance, we see the relationship of the eyes, and the eyebrows, and the nose, and all of that. We just have a pattern recognition that we're able to apply. That's another aspect of tacit knowledge.

A third aspect: if we have a lot of experience and we see things, we can sense typicality. That means we can see anomalies, and that means that we have a sense something is not right here, something doesn't feel right. And then we start to look for the specifics about what it is that's gone wrong, and that's another aspect of tacit knowledge that we depend on to alert us to possible danger.

Another aspect of tacit knowledge is our mental models of how things work. Mental models are just the stories, the frames that we have to explain causal relationships: if this happens, that will happen, and that will happen, and we build these kinds of internal representations, these mental models about how things work.

A lot of people in New York have much more sophisticated mental models about the way the financial system works than they did back in 2006. After the meltdown in 2007 and 2008, there's a much better sense of where Wall Street comes in, how it helps, how it interferes, how perverse incentives come into play. People are much more sophisticated about the interplay of those kinds of forces. They've learned about the forces, and how they connect to each other. They can't tell you, they can't draw an easy diagram, but there's a level of sophistication that many people have that they didn't have before, and that's another aspect of tacit knowledge.

Judgments based on intuition seem mysterious because intuition doesn't involve explicit knowledge. It doesn't involve declarative knowledge about facts. Therefore, we can't explicitly trace the origins of our intuitive judgments. They come from other parts of our knowing. They come from our tacit knowledge and so they feel magical. Intuition sometimes feels like we have ESP, but it isn't magical—it's really a consequence of the experience we've built up.

Moving forward from the work on decision making, a lot of our decision depends on sense making, how we size situations up. As I've been studying sense making, I've become interested in how people realize that there is a new way to size things up, how they form insights, and where insights come from. I've been looking at incidents where people generated insights in an attempt to try to see what was going on. I realize that, again, this is certainly an aspect of tacit knowledge because an insight many times will spring forward with no warning. There's no expectation, and all of a sudden you say, "Now I know what to do." So where does that sense come from, and what can get in its way? Those are the sorts of things that I've been investigating. I've been looking at different forms of these insights to get a better idea of whether it's always the same sort of process, or are there several related processes.

I'm finding it's the latter. There are several different routes for people to develop insights. A lot of the laboratory research on insights follows one of those routes: putting people in a position where they're trying to solve a puzzle and they reach an impasse, and they're stuck, and then the key for that route is to escape the fixation, to reach the insight by realizing that they're making an inappropriate assumption. So it's really looking for the assumption that isn't working for me here, trying to find what is the assumption that's fixating me so that I can get beyond it. But that's only one of the routes that we're discovering. We're finding that there are a few other routes that are also important.

As we're looking at these examples and incidents of insight, we're noticing that several of the examples involve people helping others to gain insight, and that caught my attention. How can you help other people to gain an insight? One of the ways that allows

Gary Klein

people to do that is to help them to become aware of inconsistencies in their thinking.

Let me give you an example. It's one of my favorite examples.

A friend of mine, Doug Harrington, Sr., a number of years ago was a navy pilot. He was a good navy pilot, and being able to fly a jet plane is a tough job, but he not only could fly—I mean, the navy pilots, they not only fly, but they have to be able to take off and land on an aircraft carrier, so they have to land on a ship, and the ship is bobbing up and down, and moving around, and the waves are buffeting it. It's not like a rowboat, but there's still some movement there, and you've got to land on this moving platform. Doug could do that, he was great at it. He was an instructor, and he flew F-4s. As an instructor he would teach younger pilots to land F-4s on aircraft carriers, and to fly.

Then came a point in his career when it was time for him to move to a more sophisticated, more advanced airplane, an A-6. So he learned how to fly an A-6. No problem, because he's a very skilled pilot. Now came the day when he had to do his aircraft carrier landings with his A-6. And he comes around, he's lined up to do his first landing, he's all set, and because these are so difficult, he doesn't just do the landing by himself. There is a landing signal officer on the aircraft carrier who is watching him, vectoring all the pilots, telling them what to do, and if the LSOs don't like what they see, they wave the pilot off. And so Doug is listening to the landing signal officer, he's coming in to make his landing, and the landing signal officer is telling him, "Come right, come right."

But Doug is perfectly lined up. So he does what any good navy pilot would do under those circumstances. He ignores the landing signal officer because he knows that he's got himself lined up. And the LSO, the landing signal officer, keeps saying, "Come right," and Doug doesn't really do it, and then the landing signal officer waves him off, which is weird, because he was perfectly lined up.

So now he has to go around for another try. And again, he's perfectly lined up, and the LSO is saying, "Come right," and this time he figures, "I better do it," so he comes right a little bit more, a little bit, and a little bit, not enough, he gets waved off again. Now he has to come around another go-around, and he figures, "I'm going to run out of fuel before he runs out of patience, so I better listen to him."

He tries to follow what the instructions are, and he manages to get the landing done, but he was supposed to do six landings that day, another four landings that night. He messes up all the six landings. He makes them, but they're not good landings, and at the end of the day he's told, "Doug, you just didn't do a very good job today. You're not going to do nighttime landings, that's too risky. You have to repeat your daytime landings tomorrow, and then we'll let you do the nighttime landings. But if you don't do a good enough job tomorrow, that's it. That's the end of your flying career in the navy." Doug goes into shock because everything was working well when he woke up that morning, and now his flying career may be over, and he doesn't know what happened.

His friends on the ship are there for him, and they come over to him, they're there, and they're telling him really useful things like, "Doug, you've really got to bear down tomorrow." Or, "Doug, this is important." In other words, useless advice, and they're just making him more anxious, they're driving him crazy.

At night he's ready to go to sleep, he's hoping it would be a bad dream, he'll wake up and somehow everything will work tomorrow, but he is just dazed. There's a knock on the door of his cabin and he says, "Go away," because he doesn't want to talk to anybody. It's the landing signal officer who was atypically trying to help Doug. Not that they wouldn't be helpful, but it's not their

Gary Klein

job. Their job is to just help them land, he's not supposed to be a trainer, but he's also very troubled.

So he knocks on the door, and Doug finally lets him in, and says, "I don't want you to tell me anything. People have been telling me things, it's not useful."

And the landing signal officer says, "I'm not here to tell you anything, Doug, I just want to talk to you, I just want to ask you something."

"Okay, what do you want to know?"

"I know you're a great pilot. Obviously you had trouble today. Tell me what you're trying to do."

"I'm doing what I always do, I line up the nose of the plane on a center line of the landing strip there, and I've got it perfectly lined up, and you keep telling me, 'Come right, come right.'"

"So you're flying an A-6. What did you fly before that?"

"An F-4, that's what I've been flying for years."

"In an F-4 it was either you, or you sitting behind a student. In an A-6, you're side by side, so it's not exactly the same."

"It's a foot and a half, it's not a big difference, it wouldn't make that much of a difference. Maybe two feet. It's not going to make a difference."

"Are you sure?"

"Yeah. I mean, I just line up the nose of the plane over the center line of the runway."

So the LSO says, "Let's try something."

And if you have a chance, I'd like you to try it as I'm doing this interview, if you don't mind.

He said, "Extend your arm straight out, put your thumb up. That's the nose of your airplane. Close one eye and align your thumb with a vertical line, someplace in the room. You've got your nose of the airplane lined up at the center line. Now, move your

Insight   213 head a foot and a half over," like Doug was moved over, because now he's flying an A-6. "And pull your thumb back over to that center line."

And you see what happens. It changes the whole alignment of the airplane because you're not on the center line, and it's only a foot and a half, but there's a parallax effect here that Doug wasn't thinking about. And Doug does this little demonstration, and as soon as he does it, he says,

"Ah. I'm an idiot. Obviously that's the problem."

The next day he did his six landings during the day, and he had no trouble with them, and the four landings at night, and he went on with his career.

I love that story because the LSO helped Doug achieve an insight. He tried to explain things to Doug, but that wasn't getting Doug anywhere, so he created an environment, he created an experience that allowed Doug to see the inconsistency in his beliefs, and once Doug saw the inconsistency, his mental model changed. I think helping people to arrive at insights isn't a question of pushing the insights on the people, or trying to explain it in words, as much as it is helping people to gain the experience so they can see the inconsistency for themselves, and then all of a sudden the mental model will shift naturally and easily. And to me that's a gift that good teachers have, to be able to help the people who they're trying to support. They're trying to enlighten their students or colleagues to gain those insights.

Gary Klein

11

A Sense of Cleanliness

Simone Schnall

Director, Cambridge Embodied Cognition and Emotion Laboratory; University Senior Lecturer, Department of Social and Developmental Psychology, Cambridge.

I am a social psychologist, and study judgments and decisions from the perspective that emotions, and all kinds of feelings, including physical sensations, play a really important role. For example, such simple things as a sense of cleanliness can make a difference to how people decide whether something is right or wrong. We've been looking at, in general, how people make decisions, and how they arrive at judgments. In particular we've been studying moral judgments, that is, how do people tell right from wrong?

It used to be thought for the longest time, going back for thousands of years of philosophical investigation, that people think of why a certain behavior might be wrong. They think of all the rational reasons, all the things they can come up with, they go through all the pros and cons, and then arrive at the judgment and say, "Behavior X is either wrong, or very wrong, or not so wrong, it's fine," and so on. So it used to be thought that people think long and hard, and then figure out the answer.

Now it turns out that actually this does not seem to be the case because first of all, people don't always think that much, and many thought processes are not really conscious, but rather they happen outside of consciousness. Many thoughts just happen incidentally, and people aren't even aware of them. Therefore, instead of all these sophisticated thoughts and reasons, accidental factors enter the picture, such as feelings and intuitions—for example, a sense of, "Well, I just have an intuition that this is the case," and such factors can be much more powerful than rational thought. For morality this idea first became popular in 2001 when Jonathan Haidt published his paper on the social intuitionist model, which has been a really influential idea.

To give an example, if I ask you, "How wrong is it to falsify information on your CV in order to get a better job?" you might think that you just go through a rational process, and think of the reasons why this is wrong, or perhaps why it's not so bad. But we found that when you put people in certain emotional states, for example, if you have them sit at a table that happens to be very sticky, dirty, and disgusting, then people make different decisions. If you sit at a disgusting table, or let's say you're smelling a disgusting smell in the room, then you're more likely to say that falsifying your CV in order to get a better job is really wrong compared to somebody who sits at a clean table, or somebody who doesn't have a nasty smell around them.

Similarly we find that when you give people a chance to feel very clean and pure, they decide that something like falsifying their CV is not so bad, it's proper behavior, or it's okay, it's clean. It seems like however people happen to be feeling at the moment colors their judgments about some even very fundamental decisions of whether it is right or wrong to do something. It's quite surprising that even though we like to think there are good reasons for our decisions, oftentimes there are all these random things that just happen in our lives, and that's how we decide, for example, what is moral and what is immoral.

As far as morality goes, disgust has received a lot of attention, and there has been a lot of work on it. The flip side of it is cleanliness, or being tidy, proper, clean, pure, which has been considered the absence of disgust or contamination. But there is actually more to being clean, and having things in order. On some level even cleanliness, or the desire to feel clean and pure, has a social origin in the sense that primates show social grooming: monkeys tend to get really close to each other, they pick insects off each other's fur, and it's not just useful in terms of keeping themselves clean, but it has an important social function in terms of bonding them together.

The same monkeys that pick bugs and dirt off each other's skin actually end up getting closer in the process of literally getting closer, and they become buddies, or good friends. So it seems like this behavior of keeping each other clean, or having a desire to be clean, has consequences regarding building social relationships, getting close to others, or letting somebody get close to you. There might be something really important about being clean, feeling clean, having things being proper and tidy that goes beyond just the absence of contamination or disgust.

I started my work on emotions and judgments when I collaborated with Jerry Clore and Jonathan Haidt at the University of Virginia. We got interested in moral judgment as an example of how emotions and feelings influence judgment and decision processes, and since then I've also done some work on positive moral emotions.

We test, for example, the effect of what we call "moral elevations," which is the sense of feeling uplifted and inspired when seeing somebody else do something really positive for another person. The typical film clip we use is from the *Oprah Show*, where somebody comes on the show who has been a mentor to some disadvantaged kids who grow up in some really bad parts of town, and are probably headed for a life in gang culture, but he mentors them, and becomes like a father figure.

We show participants this elevating clip, versus in other conditions we show them a neutral clip. Those who watch the Oprah clip report feeling really inspired, and uplifted, and elevated, and it gives them a warm and fuzzy feeling in the chest. But the important thing is not only this experience, but that this feeling, in turn, leads them to help others as well.

So if I have witnessed somebody who did something really wonderful, I myself also feel like I want to be a good person, and also want to help others. After watching the Oprah clip, we set up the study such that participants think there is still a second part to come in the study, but then it looks like the computer crashes, nothing works, and they're told, "You're free to go even though you were meant to stay for an hour, you can go now," after just 10 or 15 minutes. They're about to pack up their things, and then we say, "Oh, but you know what? You could really help me out by filling out these questionnaires, they are some math problems, and they're really tedious and boring, but I need some people to fill them out, and whatever you want to complete would be really helpful."

It turns out that people sit down again, and they start filling out these questionnaires, and what is really amazing is that in the moral elevation condition, people would sit down on average for about 40 minutes and just sit and complete these boring questionnaires because they feel really motivated to help another person. The experimenter clearly needed help with the surveys, and they're really happy to help. That is compared to the control condition, where they also stay a little while, maybe 15 minutes or so, but in the elevation condition, people really went out of their way to help. We have found this now in various contexts, and it's a hopeful finding that these specific moral emotions can propel people to do good things for others.

Simone Schnall

For example, if we think of how charities try to get us to contribute, they give us good arguments and reasons of how many children are starving in Africa, and the statistics, and all of that, and they try to appeal to reason. But from our finding, it looks like a more powerful way might appeal to emotion, to get people not just to think of "these poor people who are suffering," but get them to think of how wonderful they might feel themselves when they can help, and how they might inspire others to also become the benefactor of somebody in need.

Recently we also started looking at facial muscle activity in the context of morality, such as muscles involved in smiling or frowning. We used very sensitive electrodes to measure subtle muscle movement while people were considering certain moral transgressions, in other words, how much they smiled, frowned, and, importantly, how much they pulled up their nose in disgust.

We find specific patterns of muscle activity—for example, people literally frown upon certain behaviors that harm others, and they wrinkle their nose in disgust for behaviors that violate a sense of purity or cleanliness. But the important thing is that these muscle movements are predictive of people's subsequent judgment: if I frown upon you kicking a dog, the more I frown, the more wrong I will consider this transgression later on.

Disgust is a very interesting emotion because it is more physical and basic than other emotions. Originally it evolved in relation to food. Anything that doesn't smell good might be potentially harmful to you, and if you ate some rotten or contaminated food, you might get sick. It's a very adaptive, very physical emotion. But then as time went on, so the theory goes, this very basic food-related emotion became more abstract, in the sense that people also linked moral to physical disgust. Some immoral behavior might be considered disgusting, so there is a conflation of physical and spiritual or moral disgust: disgust was initially the emotion that protects us from physical contamination, but then it became the emotion that protects us from spiritual or moral contamination; it protects us from bad things, bad behaviors, and bad people.

For morality the immediate applied domain is the law, and Bobbie Spellman and I have speculated about possible implications. If you bring jurors into a courtroom and you make them wash their hands before they go in and then they get all the information about the court case, does that play a role? There are other studies in which physical manipulations change all sorts of judgments.

For example, Lawrence Williams and John Bargh showed that if you hold a cup of coffee that's hot and subsequently judge a person, you may find that person to be more warm and friendly than if you are holding a cold drink. So if you imagine a courtroom or any sort of legal situation, even if these effects are small, if you add up hand washing, holding the cup, and all kinds of factors that are, in a way, really subtle, we find that they can influence behavior. Then you just wonder, "What does it do to these really important situations where people make decisions about who gets punished or who ends up in prison?"

The question it all boils down to is whether feeling clean or dirty, or warm or cold at the moment—do these influences play an appropriate role, or is it an influence that should be ruled out? Is it something that we should ignore, or is it something we actually should pay attention to? To what extent do we want to control for these feelings and get rid of them, or rather, regulate them and say, "These must not play a role"?

To be honest, the whole idea that feelings play such a powerful role in morality is still quite new, and we are really doing the basic studies. We are doing the basic science of just looking at the really specific effects in the laboratory, and we then have to take these effects out of the laboratory and apply them.

At the end of the day, I suppose one goal could be to let people know that these are the effects that exist, and that feelings play a role. So if you pay attention to where your decisions are coming from or what might be influencing them, you might be able to control some of these effects.

Quite a bit is known about how feelings influence thoughts, and how to counteract this influence. One really influential approach by Jerry Clore and Norbert Schwarz is called the "Affect-as-Information" approach, which is the idea that whenever we have a feeling, whenever we sense a mood or an emotion, we interpret it and give it some meaning in terms of what it is relevant for. The classic study, for example, looked at how people decide whether they're satisfied with their lives, and it makes quite a bit of difference whether you ask somebody who happens to be feeling relatively happy at the moment versus somebody who is not feeling so happy—and that is actually reliably influenced by the weather.

On sunny days people feel relatively happy, and they say life is great. On rainy days they say, "Well, you know, life is kind of miserable." But if you actually remind a person of the weather and say, "Oh, isn't it a lovely day?" or "Oh, isn't it a lousy, rainy day?" then people correct for that, and on a rainy day, for example, say that, "Well, you know, life is not so bad," because they realize that their feelings play a role as far as their judgments are concerned. So oftentimes in life we feel something, and we interpret it as being meaningful.

If I feel happy and good, I take that to mean that my life is good, even though, of course, how I feel at the moment is not really that relevant for life as a whole. This idea has been taken and found to be relevant in lots of domains, and most recently the domain that we've looked at is the moral domain, as I mentioned earlier.

Oftentimes we cannot help it, so even if we know that, "Well, it's sunny, and that doesn't really mean that everything is good," or on the contrary, if it's a cloudy day, it doesn't mean that everything is awful, but it still feels like that. Feeling can be very hard to override, and that probably has to do with the fact that certain physiological processes in the body are very fundamental. We like to think that just by making an effort and exerting willpower we can override them, but these physical sensations can be really powerful and difficult to change.

I'm not sure how fruitful it is to chase after consciousness simply because doing so might lead us astray. If a lot of what we do happens outside of consciousness or for reasons that we're not aware of, then if we try to figure out consciousness it may not explain that much. More productive ways forward might involve investigating the mind from the perspective of embodied cognition. It is the idea that a lot of thoughts or cognitive processes are not just to do with what happens in the brain, but they take into account physical sensations and information from the body, and the body in relation to performing specific actions on the environment. It's all about the very basic cognitive processes that are in place because we are very similar to other nonhuman animals. It's as simple as that. We like to think of ourselves as special because we're so intelligent, but we're still very similar to other creatures in terms of how we process the world around us.

Larry Barsalou has done a lot of work to show that the language of the mind is very similar to how the body takes in information. The idea is that when we process something, when we have a sensory experience of some kind, that same type of language or that same modality of information is taken and used in abstract

Simone Schnall

thought processes. So this is where metaphors come in. Embodied metaphors are when, in a way, we take physical concepts from the world that we experience with the body, and use them when we talk about abstract things.

For example, there is a recent empirical work that shows that metaphors have an embodied spatial basis. We like to think of good things as up, and bad things as down. When we're feeling good we're standing tall, perked up, whereas if we're feeling bad, we might literally be depressed and down. Brian Meier and Michael Robinson, for example, find that when positive things are presented in an up location on a computer screen, people are faster to identify them than when they're presented in a down location. Similarly, when something negative is presented in a down location it's easier to identify. So there is a sense that we have this spatial connotation of what is good and bad, and that seems to come from our physical orientation, simply how we stand, or how we function in physical space.

It all goes back to *Metaphors We Live By*, by George Lakoff and Mark Johnson. It was all laid out there in the theoretical context of cognitive linguistics, but it was only recently that social psychologists started to empirically test these ideas, and there is now a lot of support for it.

An interesting finding that has been popping up in the literature in various forms is the idea of some sort of moral equilibrium, such that people like to be at a minimal level of feeling morally good about themselves, and once they've reached that level, they're content to stay there. So just like embodied factors such as how clean or how disgusted I feel at the moment, how morally good I feel about myself can change my judgments and behaviors. The "moral licensing effect" suggests that if I'm reminded that I'm a pretty good person, that I do good things, that I have positive qualities and so on, I'm less likely to help another person than somebody who was not previously reminded of their positive qualities, and who has some self-doubts.

Basically, once I've affirmed the fact that I'm a pretty decent person, then that's it, then I am perfectly content, whereas if I feel inferior on a moral level, then I want to help others, and do good things. It's quite powerful to think that people have this sense of a moral equilibrium that they want to reach, but then also not go beyond, or not drop below, and this may have really important implications for why people do good things, why they fail to do good things, why they do bad things, and so on. Again, it goes back to the idea that even though we think that rational considerations are behind our thoughts and actions, they might in fact be due to factors we're not even aware of.

Morality has really turned into a hot topic within the last few years in psychology. Somehow it really caught on. Why is it that this idea of an intuition-based morality has become so popular? I think there are a couple of reasons. One reason is that this approach links up well with the approach of embodied cognition, which has become influential at the same time. Another reason is because there is a growing understanding that so many things happen outside of consciousness. We are not aware of many of the things we're doing, and why we are doing them. So there is a lot of agreement that, oftentimes, we do things for no good reason whatsoever. That means that for better or worse, rational thought may not really happen that often, and it's not such a central component of human life, or of why we do things.

Simone Schnall

12

The Fourth Quadrant: A Map of the Limits of Statistics

Nassim Nicholas Taleb

Essayist; former mathematical trader; Distinguished Professor of Risk Engineering, New York University's Polytechnic Institute; author, *Fooled by Randomness*, *The Black Swan*, and *Antifragile*.

INTRODUCTION by John Brockman

When Nassim Taleb talks about the limits of statistics, he becomes outraged. "My outrage," he says, "is aimed at the scientist-charlatan putting society at risk using statistical methods. This is similar to iatrogenics, the study of the doctor putting the patient at risk." As a researcher in probability, he has some credibility. In 2006, using FNMA and bank risk managers as his prime perpetrators, he wrote the following:

> *The government-sponsored institution Fannie Mae, when I look at its risks, seems to be sitting on a barrel of dynamite, vulnerable to the slightest hiccup. But not to worry: their large staff of scientists deemed these events "unlikely."*

In the following Edge *original essay, Taleb continues his examination of black swans, the highly improbable and unpredictable events that have massive impact. He claims that those who are putting society at risk are "no true statisticians," merely people using statistics either without understanding them or in a self-serving manner. "The current subprime crisis did wonders to help me drill my point about the limits of statistically driven claims," he says.*

Taleb, looking at the cataclysmic situation facing financial institutions today, points out that "the banking system, betting against black swans, has lost over one trillion dollars (so far), more than was ever made in the history of banking."

But, as he points out, there is also good news.

We can identify where the danger zone is located, which I call "the fourth quadrant," and show it on a map with more or less clear boundaries. A map is a useful thing because you know where you are safe and where your knowledge is questionable. So I drew for the Edge *readers a tableau showing the boundaries where statistics works well and where it is questionable or unreliable. Now once you identify where the danger zone is, where your knowledge is no longer valid, you can easily make some policy rules: how to conduct yourself in that fourth quadrant; what to avoid.*

Statistical and applied probabilistic knowledge is the core of knowledge; statistics is what tells you if something is true, false, or merely anecdotal; it is the "logic of science"; it is the instrument of risk taking; it is the applied tools of epistemology; you can't be a modern intellectual and not think probabilistically—but . . . let's not be suckers. The problem is much more complicated than it seems to the casual, mechanistic user who picked it up in graduate school. Statistics can fool you. In fact it is fooling your government right now. It can even bankrupt the system (let's face it: use of probabilistic methods for the estimation of risks did just blow up the banking system).

The current subprime crisis has been doing wonders for the reception of any ideas about probability-driven claims in science, particularly in social science, economics, and "econometrics" (quantitative economics). Clearly, with current International Monetary Fund estimates of the costs of the 2007–2008 subprime crisis, the banking

Nassim Nicholas Taleb

system seems to have lost more on risk taking (from the failures of quantitative risk management) than every penny banks ever earned taking risks. But it was easy to see from the past that the pilot did not have the qualifications to fly the plane and was using the wrong navigation tools: the same happened in 1983 with money center banks losing cumulatively every penny ever made, and in 1991–1992 when the Savings and Loans industry became history.

It appears that financial institutions earn money on transactions (say fees on your mother-in-law's checking account) and lose everything taking risks they don't understand. I want this to stop, and stop now—the current patching by the banking establishment worldwide is akin to using the same doctor to cure the patient when the doctor has a track record of systematically killing them. And this is not limited to banking—I generalize to an entire class of random variables that do not have the structure we think they have, in which we can be suckers.

And we are beyond suckers: not only, for socioeconomic and other nonlinear, complicated variables, are we riding in a bus driven by a blindfolded driver, but we refuse to acknowledge it in spite of the evidence, which to me is a pathological problem with academia. After 1998, when a "Nobel-crowned" collection of people (and the crème de la crème of the financial economics establishment) blew up Long Term Capital Management, a hedge fund, because the "scientific" methods they used misestimated the role of the rare event, such methodologies and such claims on understanding risks of rare events should have been discredited. Yet the Fed helped their bailout, and exposure to rare events (and model error) patently increased exponentially (as we can see from banks' swelling portfolios of derivatives that we do not understand).

Are we using models of uncertainty to produce certainties?

This masquerade does not seem to come from statisticians—but from the commoditized, "me-too" users of the products. Professional statisticians can be remarkably introspective and self-critical. Recently, the American Statistical Association had a special panel session on the "black swan" concept at the annual Joint Statistical Meeting in Denver last August. They insistently made a distinction between the "statisticians" (those who deal with the subject itself and design the tools and methods) and those in other fields who pick up statistical tools from textbooks without really understanding them. For them it is a problem with statistical education and half-baked expertise. Alas, this category of blind users includes regulators and risk managers, whom I accuse of creating more risk than they reduce.

So the good news is that we can identify where the danger zone is located, which I call "the fourth quadrant," and show it on a map with more or less clear boundaries. A map is a useful thing because you know where you are safe and where your knowledge is questionable. So I drew for the *Edge* readers a tableau showing the boundaries where statistics works well and where it is questionable or unreliable. Now once you identify where the danger zone is, where your knowledge is no longer valid, you can easily make some policy rules: how to conduct yourself in that fourth quadrant; what to avoid.

So the principal value of the map is that it allows for policy making. Indeed, I am moving on: my new project is about methods on how to domesticate the unknown, exploit randomness, figure out how to live in a world we don't understand very well. While most human thought (particularly since the Enlightenment) has focused us on how to turn knowledge into decisions, my new mission is to build methods to turn lack of information, lack of understanding, and lack of "knowledge" into decisions—how, as we will see, not to be a "turkey."

This piece has a technical appendix that presents mathematical points and empirical evidence. (See link below.) It includes a battery of tests showing that no known conventional tool can allow us to make

Nassim Nicholas Taleb

precise statistical claims in the fourth quadrant. While in the past I limited myself to citing research papers and evidence compiled by others (a less risky trade), here I got hold of more than 20 million pieces of data (includes 98 percent of the corresponding macroeconomics values of transacted daily, weekly, and monthly variables for the last 40 years) and redid a systematic analysis that includes recent years.

What is fundamentally different about real life

My anger with "empirical" claims in risk management does not come from research. It comes from spending twenty tense (but entertaining) years taking risky decisions in the real world managing portfolios of complex derivatives, with payoffs that depend on higher-order statistical properties—and you quickly realize that a certain class of relationships that "look good" in research papers almost never replicate in real life (in spite of the papers making some claims with a "p" close to infallible). But that is not the main problem with research.

For us the world is vastly simpler in some sense than the academy, vastly more complicated in another. So the central lesson from decision making (as opposed to working with data on a computer or bickering about logical constructions) is the following: it is the exposure (or payoff) that creates the complexity—and the opportunities and dangers—not so much the knowledge (i.e., statistical distribution, model representation, etc.). In some situations, you can be extremely wrong and be fine; in others you can be slightly wrong and explode. If you are leveraged, errors blow you up; if you are not, you can enjoy life.

So knowledge (i.e., if some statement is "true" or "false") matters little, very little, in many situations. In the real world, there are very few situations where what you do and your belief if some statement is true or false naively map into each other. Some decisions require vastly more caution than others—or highly more drastic confidence intervals. For instance, you do not "need evidence" that the water is poisonous to not drink from it. You do not need "evidence" that a gun is loaded to avoid playing Russian roulette, or evidence that a thief is on the lookout to lock your door. You need evidence of safety—not evidence of lack of safety—a central asymmetry that affects us with rare events. This asymmetry in skepticism makes it easy to draw a map of danger spots.

The dangers of bogus math

I start with my old crusade against "quants" (people like me who do mathematical work in finance), economists, and bank risk managers, my prime perpetrators of iatrogenic risks (the healer killing the patient). Why iatrogenic risks? Because not only have economists been unable to prove that their models work, but no one has managed to prove that the use of a model that does not work is neutral, that it does not increase blind risk taking, hence the accumulation of hidden risks.

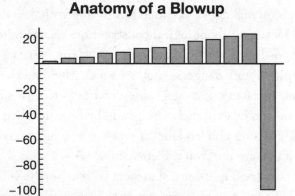

Figure 1. My classical metaphor: A turkey is fed for 1,000 days—every day confirms to its statistical department that the human race cares about its welfare "with increased statistical significance." On the 1,001st day, the turkey has a surprise.

Nassim Nicholas Taleb

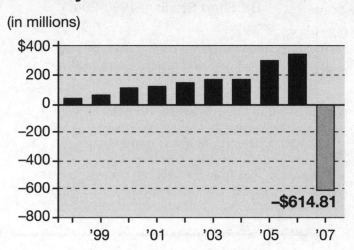

IndyMac's annual net income

(in millions)

Source: **Bloomberg News**

Figure 2. The graph above shows the fates of close to 1,000 financial institutions (including busts such as FNMA, Bear Stearns, Northern Rock, Lehman Brothers, etc.). The banking system (betting AGAINST rare events) just lost >one trillion dollars (so far) on a single error, more than was ever earned in the history of banking. Yet bankers kept their previous bonuses, and it looks like citizens have to foot the bills. And one Professor Ben Bernanke pronounced right before the blowup that we live in an era of stability and "great moderation" (he is now piloting a plane and we all are passengers on it).

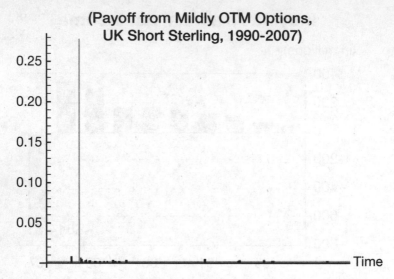

(Payoff from Mildly OTM Options, UK Short Sterling, 1990-2007)

Figure 3. The graph above shows the daily variations in a derivatives portfolio exposed to UK interest rates between 1988 and 2008. Close to 99 percent of the variations, over the span of 20 years, will be represented in one single day—the day the European Monetary System collapsed. As I show in the appendix, this is typical with ANY socioeconomic variable (commodity prices, currencies, inflation numbers, GDP, company performance, etc.). No known econometric statistical method can capture the probability of the event with any remotely acceptable accuracy (except, of course, in hindsight, and "on paper"). Also note that this applies to surges on electricity grids and all manner of modern-day phenomena.

Figures 1 and 2 show you the classical problem of the turkey making statements on risks based on past history (mixed with some theorizing that happens to narrate well with the data). A friend of mine was sold a package of subprime loans (leveraged) on grounds that "30 years of history show that the trade is safe." He found the argument unassailable "empirically." And the unusual dominance of the rare event shown in Figure 3 is not unique: it affects all macroeconomic data—if you look long enough, almost all the contribution in some classes of variables will come from rare events (I looked in the appendix at 98 percent of trade-weighted data).

Now let me tell you what worries me. Imagine that the turkey

Nassim Nicholas Taleb

can be the most powerful man in world economics, managing our economic fates. How? A then Princeton economist named Ben Bernanke made a pronouncement in late 2004 about the "new moderation" in economic life—the world getting more and more stable—before becoming the chairman of the Federal Reserve. Yet the system was getting riskier and riskier as we were turkey-style sitting on more and more barrels of dynamite—and Professor Bernanke's predecessor, the former Federal Reserve Chairman Alan Greenspan, was systematically increasing the hidden risks in the system, making us all more vulnerable to blowups.

By the "narrative fallacy," the turkey economics department will always manage to state, before Thanksgivings, that "we are in a new era of safety," and back it up with thorough and "rigorous" analysis. And Professor Bernanke indeed found plenty of economic explanations—what I call the narrative fallacy—with graphs, jargon, curves, the kind of façade-of-knowledge that you find in economics textbooks. (This is the find of glib, snake-oil façade of knowledge—even more dangerous because of the mathematics—that made me, before accepting the new position in NYU's engineering department, verify that there was not a single economist in the building. I have nothing against economists: you should let them entertain each other with their theories and elegant mathematics, and help keep college students inside buildings. But beware: they can be plain wrong, yet frame things in a way to make you feel stupid arguing with them. So make sure you do not give any of them risk-management responsibilities.)

Bottom line: the map

Things are made simple by the following. There are two distinct types of decisions, and two distinct classes of randomness.

Decisions

The first type of decision is simple, "binary," i.e., you just care if something is true or false. Very true or very false does not matter. Someone is either pregnant or not pregnant. A statement is "true" or "false" with some confidence interval. (I call these M0 as, more technically, they depend on the zeroth moment, namely just on probability of events, and not their magnitude—you just care about "raw" probability.) A biological experiment in the laboratory or a bet with a friend about the outcome of a soccer game belongs to this category.

The second type of decision is more complex. You do not just care about the frequency—but of the impact as well, or, even more complex, some function of the impact. So there is another layer of uncertainty of impact. (I call these M1+, as they depend on higher moments of the distribution.) When you invest you do not care how many times you make or lose, you care about the expectation: how many times you make or lose times the amount made or lost.

Probability structures

There are two classes of probability domains—very distinct qualitatively and quantitatively. The first, thin-tailed Mediocristan; the second, thick-tailed Extremistan. Before I get into the details, take the literary distinction as follows:

> In Mediocristan, exceptions occur but don't carry large consequences. Add the heaviest person on the planet to a sample of 1,000. The total weight would barely change. In Extremistan, exceptions can be everything (they will eventually, in time, represent everything). Add Bill Gates to your sample: the wealth will jump by a factor of >100,000. So,

in Mediocristan, large deviations occur but they are not consequential—unlike in Extremistan.

Mediocristan corresponds to "random walk"-style randomness that you tend to find in regular textbooks (and in popular books on randomness). Extremistan corresponds to a "random jump" one. The first kind I can call "Gaussian-Poisson," the second "fractal" or Mandelbrotian (after the works of the great Benoit Mandelbrot, linking it to the geometry of nature). But note here an epistemological question: there is a category of "I don't know" that I also bundle in Extremistan for the sake of decision making—simply because I don't know much about the probabilistic structure or the role of large events.

The map

Now let's see where the traps are:

First quadrant
Simple binary decisions, in Mediocristan: statistics does wonders. These situations are, unfortunately, more common in academia, laboratories, and games than real life—what I call the "ludic fallacy." In other words, these are the situations in casinos, games, and dice, and we tend to study them because we are successful in modeling them.

Second quadrant
Simple decisions, in Extremistan: some well-known problem studied in the literature. Except of course that there are not many simple decisions in Extremistan.

Third quadrant
Complex decisions in Mediocristan: statistical methods work surprisingly well.

Fourth quadrant
Complex decisions in Extremistan: welcome to the black swan domain. Here is where your limits are. Do not base your decisions on statistically based claims. Or, alternatively, try to move your exposure type to make it third-quadrant style ("clipping tails").

APPLICATION	Simple payoffs	Complex payoffs
DOMAIN		
Distribution 1 ("thin tailed")	Extremely robust to Black Swans	Quite robust to Black Swans
Distribution 2 ("heavy" and/or unknown tails, no or unknown characteristic scale)	Quite robust to Black Swans	LIMITS of Statistics – extreme fragility to Black Swans

The four quadrants. The southeast area is where statistics and models fail us.

Nassim Nicholas Taleb

Tableau of payoffs

SIMPLE PAYOFFS "True/False?"	COMPLEX PAYOFFS "How Much?"	VERY COMPLEX PAYOFFS "How Much? Really?"
M0 (depend on raw probability, i.e. probability times payoff to the 0th power)	**M1** Expectations LINEAR PAYOFF (Simple expectation, probability times payoff)	**M2+** NONLINEAR PAYOFF (Probability times a power (or convex function) of the payoff– square, cubic, etc.)
Medicine (health not epidemics)	Security: Terrorism, Natural catastrophes	Payoffs from innovation, convex technologies
Psychological experiments	Environmental problems	Societal consequences of pandemics
Bets (prediction markets)	Climate	Calibration of nonlinear models
Binary/Digital derivatives	General risk management	Expectation weighted by nonlinear utility
Life/Death (for one person)	Epidemics, Life/Death for n persons (n stochastic)	Errors in analyses of deviations
Bankruptcy (for a single company)	Insurance, measures of expected shortfall	Leveraged portfolios (around the loss point)
Single insurance (capped)	Insurance (large portfolio), open-ended catastrophe insurance	Derivative payoffs, Dynamically hedged portfolios
What Else? **Very Little!**	Finance (Investments)	Kurtosis-based positioning ("volatility trading")
	Casinos	Cubic payoffs (strips of out of the money options)
	Economics (Policy)	

Two difficulties

Let me refine the analysis. The passage from theory to the real world presents two distinct difficulties: "inverse problems" and "preasymptotics."

Inverse problems

It is the greatest epistemological difficulty I know. In real life we do not observe probability distributions (not even in Soviet Russia, not even the French government). We just observe events. So we do not know the statistical properties—until, of course, after the fact. Given a set of observations, plenty of statistical distributions can correspond to the exact same realizations—each would extrapolate differently outside the set of events on which it was derived. The inverse problem is more acute when more theories, more distributions can fit a set of data.

This inverse problem is compounded by the small sample properties of rare events, as these will be naturally rare in a past sample. It is also acute in the presence of nonlinearities as the families of possible models/parametrization explode in numbers.

Preasymptotics

Theories are, of course, bad, but they can be worse in some situations when they were derived in idealized situations, the asymptote, but are used outside the asymptote (its limit, say infinity or the infinitesimal). Some asymptotic properties do work well preasymptotically (Mediocristan), which is why casinos do well, but others do not, particularly when it comes to Extremistan.

Most statistical education is based on these asymptotic, Platonic properties—yet we live in the real world that rarely resembles the asymptote. Furthermore, this compounds the ludic fallacy: most of what students of statistics do is assume a structure, typically with a known probability. Yet the problem we have is not so much making computations once you know the probabilities, but finding the true distribution.

The inverse problem of the rare events

Let us start with the inverse problem of rare events and proceed with a simple, nonmathematical argument. In August 2007, the *Wall Street Journal* published a statement by one financial economist, expressing his surprise that financial markets experienced a string of events that "would happen once in 10,000 years." A portrait of the gentleman accompanying the article revealed that he was considerably younger than 10,000 years; it is therefore fair to assume that he was not drawing his inference from his own empirical experience (and not from history at large), but from some theoretical model that produces the risk of rare events, or what he perceived to be rare events.

Alas, the rarer the event, the more theory you need (since we don't observe it). So the rarer the event, the worse its inverse problem. And theories are fragile (just think of Professor Bernanke).

The tragedy is as follows. Suppose that you are deriving probabilities of future occurrences from the data, assuming (generously) that the past is representative of the future. Now, say that you estimate that an event happens every 1,000 days. You will need a lot more data than 1,000 days to ascertain its frequency, say 3,000 days. Now, what if the event happens once every 5,000 days? The estimation of this probability requires some larger number, 15,000 or more. The smaller the probability, the more observations you need, and the greater the estimation error for a set number of observations. Therefore, to estimate a rare event you need a sample that is larger and larger in inverse proportion to the occurrence of the event.

If small probability events carry large impacts, and (at the same time) these small probability events are more difficult to compute from past data itself, then: our empirical knowledge about the potential contribution—or role—of rare events (probability × consequence) is inversely proportional to their impact. This is why we should worry in the fourth quadrant!

For rare events, the confirmation bias (the tendency, Bernanke-style, of finding samples that confirm your opinion, not those that disconfirm it) is very costly and very distorting. Why? Most of histories of black swan–prone events is going to be black swan–free! Most samples will not reveal the black swans—except after if you are hit with them, in which case you will not be in a position to discuss them. Indeed I show with 40 years of data that past black swans do not predict future black swans in socioeconomic life.

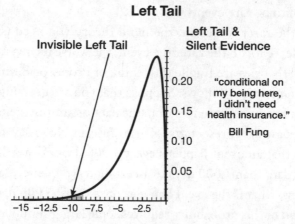

Figure 4. The Confirmation Bias at Work. For left-tailed fat-tailed distributions, we do not see much of negative outcomes for surviving entities AND we have a small sample in the left tail. This is why we tend to see a better past for a certain class of time series than warranted.

Nassim Nicholas Taleb

Fallacy of the single event probability

Let us look at events in Mediocristan. In a developed country a newborn female is expected to die at around 79, according to insurance tables. When she reaches her 79th birthday, her life expectancy, assuming that she is in typical health, is another 10 years. At the age of 90, she should have another 4.7 years to go. So if you are told that a person is older than 100, you can estimate that he is 102.5, and conditional on the person being older than 140 you can estimate that he is 140 plus a few minutes. The conditional expectation of additional life drops as a person gets older.

In Extremistan things work differently and the conditional expectation of an increase in a random variable does not drop as the variable gets larger. In the real world, say with stock returns (and all economic variables), conditional on a loss being worse than the 5 units, to use a conventional unit of measure units, it will be around 8 units. Conditional that a move is more than 50 STD it should be around 80 units, and if we go all the way until the sample is depleted, the average move worse than 100 units is 250 units! This extends all the way to areas in which we have sufficient samples.

This tells us that there is "no typical" failure and "no typical" success. You may be able to predict the occurrence of a war, but you will not be able to gauge its effect! Conditional on a war killing more than 5 million people, it should kill around 10 (or more). Conditional on it killing more than 500 million, it would kill a billion (or more, we don't know). You may correctly predict a skilled person getting "rich," but he can make a million, ten million, a billion, ten billion—there is no typical number. We have data, for instance, for predictions of drug sales, conditional on getting things right. Sales estimates are totally uncorrelated to actual sales—some drugs that were correctly predicted to be successful had their sales underestimated by up to 22 times!

This absence of "typical" events in Extremistan is what makes prediction markets ludicrous, as they make events look binary. "A war" is meaningless: you need to estimate its damage—and no damage is typical. Many predicted that the First World War would occur—but nobody predicted its magnitude. One of the reasons economics does not work is that the literature is almost completely blind to the point.

A simple proof of unpredictability in the fourth quadrant

I show elsewhere that if you don't know what a "typical" event is, fractal power laws are the most effective way to discuss the extremes mathematically. It does not mean that the real-world generator is actually a power law—it means you don't understand the structure of the external events it delivers and need a tool of analysis so you do not become a turkey. Also, fractals simplify the mathematical discussions because all you need is play with one parameter (I call it "alpha") and it increases or decreases the role of the rare event in the total properties.

For instance, you move alpha from 2.3 to 2 in the publishing business, and the sales of books in excess of 1 million copies triple! Before meeting Benoit Mandelbrot, I used to play with combinations of scenarios with series of probabilities and series of payoffs filling spreadsheets with clumsy simulations; learning to use fractals made such analyses immediate. Now all I do is change the alpha and see what's going on.

Now the problem: parametrizing a power law lends itself to monstrous estimation errors (I said that heavy tails have horrible inverse problems). Small changes in the "alpha" main parameter used by power laws lead to monstrously large effects in the tails. Monstrous.

And we don't observe the "alpha." Figure 5 shows more than 40,000 computations of the tail exponent "alpha" from different samples of different economic variables (data for which it is impossible to refute fractal power laws). We clearly have problems figuring out what the "alpha" is: our results are marred with errors. Clearly the mean absolute error is in excess of 1 (i.e., between alpha=2 and alpha=3). Numerous papers in econophysics found an "average" alpha between 2 and 3—but if you process the >20 million pieces of data analyzed in the literature, you find that the variations between single variables are extremely significant.

Now this mean error has massive consequences. Figure 6 shows the effect: the expected value of your losses in excess of a certain amount (called "shortfall") is multiplied by >10 from a small change in the "alpha" that is less than its mean error! These are the losses banks were talking about with confident precision!

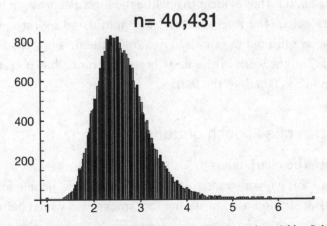

Figure 5. Estimation error in "alpha" from 40,000 economic variables. I thank Pallop Angsupun for the data.

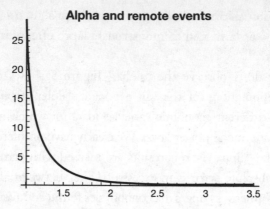

Figure 6. The value of the expected shortfall (expected losses in excess of a certain threshold) in response to changes in tail exponent "alpha." We can see it explode by an order of magnitude.

What if the distribution is not a power law? This is a question I used to get once a day. Let me repeat it: my argument would not change—it would take longer to phrase it.

Many researchers, such as Philip Tetlock, have looked into the incapacity of social scientists in forecasting (economists, political scientists). It is thus evident that while the forecasters might be just "empty suits," the forecast errors are dominated by rare events, and we are limited in our ability to track them. The "wisdom of crowds" might work in the first three quadrants; but it certainly fails (and has failed) in the fourth.

Living in the fourth quadrant

Beware the charlatan

When I was a quant-trader in complex derivatives, people mistaking my profession used to ask me for "stock tips," which put me in a state of rage: a charlatan is someone likely (statistically) to give you positive advice, of the "how to" variety.

Go to a bookstore, and look at the business shelves: you will find plenty of books telling you how to make your first million, or your first quarter-billion, etc. You will not be likely to find a book on "how I failed in business and in life"—though the second type of advice is vastly more informational, and typically less charlatanic. Indeed, the only popular such finance book I found that was not quacky in nature—on how someone lost his fortune—was both self-published and out of print. Even in academia, there is little room for promotion by publishing negative results—though these are vastly informational and less marred with statistical biases of the kind we call data snooping. So all I am saying is "what is it that we don't know," and my advice is what to avoid, no more.

You can live longer if you avoid death, get better if you avoid bankruptcy, and become prosperous if you avoid blowups in the fourth quadrant.

Now you would think that people would buy my arguments about lack of knowledge and accept unpredictability. But many kept asking me, "Now that you say that our measures are wrong, do you have anything better?"

I used to give the same mathematical finance lectures for both graduate students and practitioners before giving up on academic students and grade-seekers. Students cannot understand the value of "this is what we don't know"—they think it is not information, that they are learning nothing. Practitioners, on the other hand, value it immensely. Likewise with statisticians: I never had a disagreement with statisticians (who build the field)—only with users of statistical methods.

Spyros Makridakis and I are editors of a special issue of a decision science journal, *The International Journal of Forecasting*. The issue is about "what to do in an environment of low predictability." We received tons of papers, but guess what? Very few addressed the point: they mostly focused on showing us that they predict better (on paper). This convinced me to engage in my new project: "how to live in a world we don't understand."

So for now I can produce phronetic rules (in the Aristotelian sense of phronesis, decision-making wisdom). Here are a few, to conclude.

Phronetic rules: what is wise to do (or not do) in the fourth quadrant

1) Avoid optimization, learn to love redundancy.
Psychologists tell us that getting rich does not bring happiness—if you spend it. But if you hide it under the mattress, you are less vulnerable to a black swan. Only fools (such as banks) optimize, not realizing that a simple model error can blow through their capital (as it just did). In one day in August 2007, Goldman Sachs experienced 24 times the average daily transaction volume—would 29 times have blown up the system? The only weak point I know of financial markets is their ability to drive people and companies to "efficiency" (to please a stock analyst's earnings target) against risks of extreme events.

Indeed, some systems tend to optimize—and therefore become more fragile. Electricity grids, for example, optimize to the point of not coping with unexpected surges—Albert-László Barabási warned us of the possibility of an NYC blackout like the one we had in August 2003. Quite prophetic, the fellow. Yet energy supply has kept getting more and more efficient since. Commodity prices can double on a short burst in demand (oil, copper, wheat)—we no longer have any slack. Almost everyone who talks about "flat earth" does not realize that it is overoptimized to the point of maximal vulnerability.

Biological systems—those that survived millions of years—

Nassim Nicholas Taleb

include huge redundancies. Just consider why we like sexual encounters (so redundant to do it so often!). Historically populations tended to produced around 4 to 12 children to get to the historical average of ~2 survivors to adulthood.

Option-theoretic analysis: redundancy is like long an option. You certainly pay for it, but it may be necessary for survival.

2) Avoid prediction of remote payoffs

—though not necessarily ordinary ones. Payoffs from remote parts of the distribution are more difficult to predict than closer parts.

A general principle is that while in the first three quadrants you can use the best model you can find, this is dangerous in the fourth quadrant: no model should be better than just any model.

3) Beware the "atypicality" of remote events.

There is a sucker's method called "scenario analysis" and "stress testing"—usually based on the past (or some "make sense" theory). Yet I show in the appendix how past shortfalls do not predict subsequent shortfalls. Likewise, "prediction markets" are for fools. They might work for a binary election, but not in the fourth quadrant. Recall that the very definition of events is complicated: success might mean one million in the bank . . . or five billion!

4) Time.

It takes much, much longer for a times series in the fourth quadrant to reveal its property. At the worst, we don't know how long. Yet compensation for bank executives is done on a short-term window, causing a mismatch between observation window and necessary window. They get rich in spite of negative returns. But we can have a pretty clear idea if the black swan can hit on the left (losses) or on the right (profits).

The point can be used in climatic analysis. Things that have worked for a long time are preferable—they are more likely to have reached their ergodic states.

5) Beware moral hazard.

It is optimal to make a series of bonuses betting on hidden risks in the fourth quadrant, then blow up and write a thank-you letter. Fannie Mae and Freddie Mac's chairmen will in all likelihood keep their previous bonuses (as in all previous cases) and even get close to $15 million of severance pay each.

6) Metrics.

Conventional metrics based on type 1 randomness don't work. Words like "standard deviation" are not stable and do not measure anything in the fourth quadrant. So does "linear regression" (the errors are in the fourth quadrant), "Sharpe ratio," Markowitz optimal portfolio, ANOVA shmnamova, least square, etc. Literally anything mechanistically pulled out of a statistical textbook.

My problem is that people can both accept the role of rare events, agree with me, *and* still use these metrics, which is leading me to test whether this is a psychological disorder.

The technical appendix shows why these metrics fail: they are based on "variance"/"standard deviation" and terms invented years ago when we had no computers. One way I can prove that anything linked to standard deviation is a façade of knowledge: there is a measure called Kurtosis that indicates departure from "Normality." It is very, very unstable and marred with huge sampling error: 70 to 90 percent of the Kurtosis in Oil, SP500, silver, UK interest rates, Nikkei, U.S. deposit rates, sugar, and the dollar/yen currency rate come from one day in the past 40 years, reminiscent of Figure 3. This means that no sample will ever deliver the true variance. It also tells us that anyone using "variance" or "standard deviation" (or worse, making models that make us take decisions based on it) in the fourth quadrant is incompetent.

7) Where is the skewness?

Clearly the fourth quadrant can present left or right skewness. If we suspect right skewness, the true mean is more likely to be underestimated by measurement of past realizations, and the total potential is likewise poorly gauged. A biotech company (usually) faces positive uncertainty, a bank faces almost exclusively negative shocks. I call that in my new project "concave" or "convex" to model error.

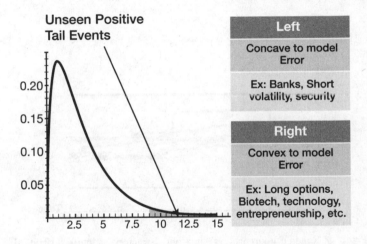

Unseen Positive Tail Events

Left	
Concave to model Error	
Ex: Banks, Short volatility, security	

Right	
Convex to model Error	
Ex: Long options, Biotech, technology, entrepreneurship, etc.	

8) Do not confuse absence of volatility with absence of risks.

Recall how conventional metrics of using volatility as an indicator of stability have fooled Bernanke—as well as the banking system.

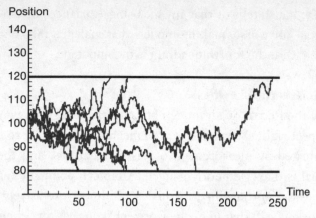

Figure 7. Random Walk—characterized by volatility. You only find these in textbooks and in essays on probability by people who have never really taken decisions under uncertainty.

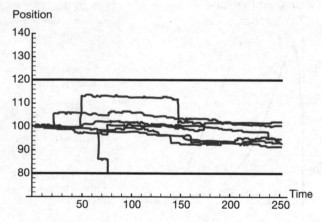

Figure 8. Random Jump process—it is not characterized by its volatility. It exits the 80–120 range much less often, but its extremes are far more severe. Please tell Bernanke if you have the chance to meet him.

9) Beware presentations of risk numbers.

Not only do we have mathematical problems, but risk perception is subjected to framing issues that are acute in the fourth quadrant. Dan Goldstein and I are running a program of experiments in the psychology of uncertainty and finding that the perception of rare events is subjected to severe framing distortions: people are aggressive with risks that hit them "once every 30 years" but not if they are told that the risk happens with a "3 percent a year" occurrence. Furthermore, it appears that risk representations are not neutral: they cause risk taking even when they are known to be unreliable.

The technical appendix to this essay can be found at http://www.fooledbyrandomness.com/EDGE/index.html.

13
Life Is the Way the Animal Is in the World

Alva Noë

Professor of Philosophy, University of California–Berkeley;
author, *Vision and Mind, Action in Perception*, and *Out of Our Heads*.

The central thing that I think about is our nature, our human-animal nature, our being in this world. What is a person? What is a human being? What is consciousness? There is a tremendous amount of enthusiasm at the moment about these questions.

They are usually framed as questions about the brain, about how the brain makes consciousness happen, how the brain constitutes who we are, what we are, what we want—our behavior. The thing I find so striking is that, at the present time, we actually can't give any satisfactory explanations about the nature of human experience in terms of the functioning of the brain.

What explains this is really quite simple. You are not your brain. You have a brain, yes. But you are a living being that is connected to an environment; you are embodied, and dynamically interacting with the world. We can't explain consciousness in terms of the brain alone because consciousness doesn't happen in the brain alone.

In many ways, the new thinking about consciousness and the brain is really just the old-fashioned style of traditional philosophical thinking about these questions but presented in a new, neuroscience package. People interested in consciousness have tended to make certain assumptions, take certain things for granted. They take for granted that thinking, feeling, wanting, consciousness in general is something that happens inside of us. They take for granted that the world, and the rest of our body, matters for consciousness only as a source of causal impingement on what is happening inside of us. Action has no more intimate connection to thought, feeling, consciousness, and experience. They tend to assume that we are fundamentally intellectual—that the thing inside of us which thinks and feels and decides is, in its basic nature, a problem solver, a calculator, a something whose nature is to figure out what there is and what we ought to do in light of what is coming in.

We should reject the idea that the mind is something inside of us that is basically matter of just a calculating machine. There are different reasons to reject this. But one is, simply put: there is nothing inside us that thinks and feels and is conscious. Consciousness is not something that happens in us. It is something we do.

A much better image is that of the dancer. A dancer is locked into an environment, responsive to music, responsive to a partner. The idea that the dance is a state of us, inside of us, or something that happens in us is crazy. Our ability to dance depends on all sorts of things going on inside of us, but that we are dancing is fundamentally an attunement to the world around us.

And this idea that human consciousness is something we enact or achieve, in motion, as a way of being part of a larger process, is the focus of my work.

Experience is something that is temporarily extended and active. Perceptual consciousness is a style of access to the world around us. I can touch something, and when I touch something I make use of an understanding of the way in which my own movements help me secure access to that which is before me. The point is not merely that I learn about or achieve access to the world by touching. The point is that the thing shows up for me as something in a space of movement-oriented possibilities.

Visual consciousness relies on a whole set of practical skills that we have, making use of the eyes and the head. I understand that if I move my eyes, I produce a certain kind of sensory change. Perceptual consciousness is a mode of exploration of the world, making use of a certain kind of practical bodily understanding. And that is what dance is. And this makes dance, for me, the perfect metaphor for consciousness.

But there's more to the comparison with dance.

Consider this. On the traditional conception of the mind, if you want to study experience, you shut your eyes and you introspect. You look inward and reflect on what is going on inside of you, on the inner show. But if experience, if seeing, hearing, thinking, and feeling, isn't something going on inside of you, but is something you do, then you need a different paradigm of what phenomenology would be, that is, of what a reflection on experience itself would be.

To reflect on experience is not to look inward; it is to pay attention to what you are doing, and to the way in which what you are doing is world- and situation- and environment-involving. Suppose I am a hiker. I walk along and move my legs in all sorts of subtle ways to follow a path along a trail. But the steps I take and the way I move my legs are modulated by, controlled by, the textures and bumps and patterns of the trail itself. There is a kind of locking in. To study experience, to think about the nature of experience, is to look at this two-way dynamic exchange between world and the active perceiver.

Not only is dance a good analogy for what consciousness is, but the experience of watching dance and the way in which we

Alva Noë

can cultivate our aesthetic appreciation of something like dance is, actually, a good way of thinking about what phenomenology itself could be. What do you see when you look at a dance? You understand the movements and the forms and the patterns of the ensemble in a particular dance environment, which may be a stage or it may be some other kind of environment. To watch a dance is to make sense of this kind of dynamic.

Contemporary dance—contemporary art more generally—can be hard to appreciate. If you're not already familiar with an artist's work, it can be difficult even to bring it into focus. But we do. It is interesting to compare this process whereby we bring a dance or other work of art into focus for aesthetic experience with the project of phenomenology itself, that is, with the project of bringing experience into focus for science. Scientists ask, how does our biological being enable us to have the kinds of experiences we have? That should be understood as a question less about how the function of our brain produces images inside our skull and, rather, about how our full embodiment enables us to carry on as we do in an environment in a situation. This raises an interesting possibility. Maybe we can think of aesthetic experience as a model of the workings at least of an important core of human consciousness—perceptual consciousness. And then maybe we can think of artistic, creative, aesthetic practice as making a direct contribution to the study of mind itself. Art is not something for science to explain; art is a domain for scientific investigation, a potential collaborator for science. It is certainly clear that the empirical investigation of consciousness requires help framing the phenomena of interest for itself.

One experience that I've been especially interested in is our understanding and experience of pictures. If I show you a picture from a newspaper—for example, a photo of Hillary Clinton—

there is a sense in which, when you look at that picture, you see Hillary. There she is, in the picture. Of course, Hillary is not there, so there is an obvious sense in which you don't see Hillary when you look at the picture. There is a sense in which you see her; and a sense in which you don't. She shows up for you, in the picture, even though she is not there. She shows up as not there. Getting clear about this phenomenon is the central empirical and conceptual problem about depiction.

One idea might be to say, well, seeing a picture of Hillary is just like seeing Hillary. Seeing a picture of Hillary produces in you, the perceiver, just the same effects that actually seeing Hillary would produce. The problem with that suggestion is that if that's right then we lose our sense of the difference between seeing Hillary and seeing a picture of Hillary. The distinctive thing about seeing Hillary in a picture is that she is there but not there. She is there but visually absent. She is manifestly absent in her visual presence. It's a kind of a paradoxical thing. There is something paradoxical about pictures.

My view is that traditional philosophy and cognitive science have been asking the wrong question when it comes to pictures. They ask, how does the picture affect us and give rise to an experience in our heads? Instead, what they should ask is, how do we achieve a kind of access to Hillary, to properties of Hillary, such as her visual appearance, by exploration of something which is not Hillary, namely, a picture?

The critical thing is the relation between this model, this picture, and that which is absent, such that we can gain access to what is absent in the picture. Once again we are thrown back to this idea that the perceiving is an achievement of access by making use of skills and knowledge. I need to know what Hillary looks like in order to recognize Hillary in her picture.

Alva Noë

A striking feature of pictures is their immediacy. A picture of Hillary doesn't seem to be a symbolic representation of Hillary. There seems to be the sense in which merely knowing how to recognize Hillary or how to recognize a human form, a figure, is enough to recognize a picture of Hillary. There is this idea that we don't need any further knowledge or further skills in order to perceive something in the picture.

That is a very interesting idea. But, in fact, there is a nice comparison we can make to help us see that pictures don't really have this sort of immediacy. Think about something like the Macintosh operating system. No promotional endorsement intended, but the Mac OS is user friendly. If you understand a few basic metaphors, about the desktop, clicking, opening files, closing files, a few basic metaphors allow you to unpack just about any program that you might be working with.

So there is a sense in which the functionality of the graphical user interface is straightforward and immediate. But, of course, that is precisely because the engineers have built the program with our particular predilections and capacities in mind. They built it to be easy for us. It's not as though it just happens to be easy. Technological evolution made it transparent for us. And pictures are just the same. You encounter pictures in a newspaper, say, and we find it easy to see Hillary Clinton in the picture. We don't need any further training. But that is not because you don't require training to see Hillary Clinton in a picture. It's because that technology was devised to be easy for us. The technology was designed for people with the training we already had.

Okay, what does that mean? Pictorial technologies, both painting and photography, have been designed to be straightforward for people who already know how to recognize things by using their eyes. Certain background visual skills are all that is presupposed.

But then seeing itself requires tremendous background knowledge.

If I have never seen a camera before, I won't know how to make sense of what that is. A beautiful paradigm for how much seeing requires background knowledge comes from art again. When you go to a museum you can look at a picture on the wall and it can be flat and unavailable and opaque. You look about it, you think about it, you talk about it, you read the placard on the wall and discuss it with a friend, and all of a sudden it can come into focus as an object. As you learn about it, you bring it visually into focus as an object. Your understanding, your thinking, helps make it intelligible. Ask this question: do you need to learn to see in pictures? Do you need to learn to see your father in a photograph?

I had an interesting example of this the other day. When I go to the museum, I often take photos with my cell phone as a record of the pictures I looked at and thought about. My son was looking through my camera and he came across this odd picture of a Dürer painting. I was in Vienna, and it was a painting that was covered with glass so that my face was also reflected in the picture of the painting. He said, "What's this?" And I said, "You tell me. What do you see?" He said, "That looks like George Washington," pointing to the businessman depicted by Dürer. "This other person, that looks like Martin Luther King." He failed to see me in the picture. He thought I was Martin Luther King.

I thought that the part that was interesting about that—my son is only six, I should add—is that so little of the stage-setting that normally goes into looking at and evaluating a photographic image was in place for him that it really was strange. The image confronted him as strange. Most of the time when we look at pictures, we do so in a context, in a setting. We can already presuppose what we are looking for, what we are interested in. These are people, these are celebrities, there are artifacts, and these are works of art.

That opens the door to the question about works of art because what makes a picture distinctively a work of art is precisely that that background presupposition is not clear, it's all in play.

The division between philosophical and empirical approaches to these questions of consciousness, understanding, and experience is an artificial one. People interested in the mind have a set of questions that they want to understand: what is thought, what is emotion, what is consciousness, what is cognition? How is it that we are able, as the animals we are, to do all this? Philosophy and science have been working on this together.

In fact, most of the science grows out of philosophical discussions. It is sometimes said by scientists that now that we have the new technologies of brain science we no longer need to pay attention to what philosophy has to say about these questions. But in fact—and this is just plain truth—most of what empirical science has to say about consciousness, language, memory, perception, and emotion is the expression of a philosophy. It comes out of an investment in a particular philosophy, namely the philosophy of the internal, the philosophy of the individual: the mind is something inside each individual; it is disconnected from other people and from the body and from the outer world. If natural science is to gain a foothold in this area, if our own nature is to become subject of empirical science, it can only be because the conceptual, methodological, philosophical, and empirical questions are approached in a new, open-minded way.

Scientists ask, what is it about the way these cells are firing in the brain that makes the corresponding experience a visual experience? It's a trick question because there is nothing about the way those cells are firing that can explain that. Certainly we don't now know anything that would allow us to point to the intrinsic properties of the cells and say, it's something about the intrinsic behavior of these cells that makes the resulting experience, the smell of coffee on a rainy morning or the redness of red. Nor can we say that populations of cells give you the solution.

We have to get bigger than that. It's not one cell; it's not populations of cells. We need to look at the whole animal's involvement with a situation. The thing about a smell is that a smell gives you the space of possible movement-sensitive changes. If I am smelling something, the movements of my nostrils in relation to the source of the odor will produce changes in the character of the odor. If we want to ask what it is about this cellular activity that makes it olfactory cellular activity, the answer is going to be the way in which the cellular activity varies as a function of the animal's movement.

And that is what the brain is doing. The brain is enabling us to establish this kind of sensorimotor engagement with the world around us. This is a substantive empirical hypothesis that I am putting forward. There are profound philosophical reasons to embrace it. And I hope that scientists and philosophers will find ways of communicating so they can work on these questions together.

Even though I'm a theoretician and not an experimentalist, philosophical research is empirically significant and I would hope that my theoretical work will contribute to the framing of theories that are empirically testable. I have collaborated with empirical researchers, although never experimentally.

In one article that I wrote with the philosopher Susan Hurley, who died in the summer of 2007, we actually made some predictions that turned out subsequently to be clinically demonstrated. In particular, we offered an account of phantom-limb pain that predicted what has subsequently been reported by V. S. Ramachandran, namely, that the use of mirrors to create sensory feedback could provide a therapy for phantom-limb pain. What

Alva Noë

Ramachandran and others have done is allow somebody who experiences phantom-limb discomfort to look at a mirror and move his good arm but get visual feedback as if he is moving the bad arm. They find that through moving the good arm it's possible to work out a cramp in what is in fact an absent arm. One of the problems of phantom-limb pain is that you can't massage it because there isn't actually a limb for you to touch. You can't work out the cramp. The sort of sensorimotor, dynamic approach that I have developed with collaborators actually predicted what they found. So that's an example of a philosophically informed empirical prediction.

I started out in the mid-to-late nineties working on visual perception. I was intrigued by the fact that there was relatively little work done on the importance of action for visual perception. The assumption was that our visual system is kind of like a camera. Action allows you to point the camera over there, but then everything just happens inside the mechanism, between your eyes and your brain.

The standard approaches that have developed over the last 100 years or so, many of which are fantastically ingenious and rich, have tended to think of vision that way: it is something that happens in the brain once the eyes get stimulated. There is one exception historically to this standard approach, a very striking exception, and that was the American psychologist J. J. Gibson. Gibson, interestingly, was a very philosophically savvy psychologist, somebody whose writing bears marks of the influence of Aristotle, Wittgenstein, and maybe even Merleau-Ponty. I view Gibson as a very important forerunner to the kind of work that I and others have been doing.

Part of the project for me has been to explore the way in which we go astray if we think of perception and action as divorced. Susan Hurley had a beautiful phrase for this. She talked about the "input-output picture," where the idea is that on this picture, perception is input from the world to the mind, action is output from the mind to the world, and cognition and consciousness is what happens inside the head to relate those two. In my view, this is all wrong. We need to get rid of the input-output picture altogether. This is what I argued in my 2004 book, *Action in Perception*. To see is to attain a certain kind of skillful access to the world. It is, for that reason, an essentially action-dependent kind of thing—by which I don't merely mean that we need to move in order to see, but by which I mean that in order to see, we need to understand what happens to us visually when we move—seeing is a kind of knowledge of the sensory effects of movement.

If I approach an object, it looms in my visual field. If I blink, the sensory stimulation from the object is disrupted. If I walk around an object, its profile transforms. In these and other ways, movement produces sensory change. I hold that seeing just is an activity of exploring the world, making use of that kind of sensorimotor understanding. The world—three-dimensional objects arrayed in space, colors, shapes, etc.—only comes into focus for perception given the perceiver's ability to exercise this kind of practical sensorimotor understanding.

There are very straightforward ways of testing this. If I put on left-right reversing goggles, you might think that what happens is that things on the left look as through they are on the right and things on the right look as though they are on the left. In fact, that's not at all what happens. If I give you descriptions written by subjects of what it is like to put on left-right reversing goggles, what they described are strange, trippy, nearly hallucinogenic experiences of boundaries between objects disintegrating, and bulges and distortions, and seemingly random movement—a breakdown of the visual world.

By hypothesis, these goggles are not distorting information. They are simply inverting it in a certain way. So why should that kind of mere inversion produce that kind of radical distortion of the character of our experience? The answer is very simple, as the psychologist Kevin O'Regan and I first showed in our 2001 *Behavioral and Brain Sciences* article on visual consciousness. When you put on those goggles, you change radically the sensory effects of your own movement. Now when you move your eyes to the left or right, you have unexpected, unanticipated consequences, and the result is not inversion but a kind of swirling, sensory confusion. If you wear the goggles long enough, it's possible to adapt to them and to see things as they are.

What explains this is the fact that through exploration of the world with the goggles one learns the new patterns of regularity, the new ways in which movement produces sensory change, the new ways in which sensory change varies systematically with movement. Once you figure out the new laws of sensory motor contingency, the world comes back into focus. What is interesting in this story is that you don't explain how the experience changed by looking at cells or populations of cells. You explain how experience changes by looking at the way in which cells function as part of a larger dynamic of activity: animal, world, brain working together to make consciousness happen.

One of the key thoughts here, then, is that if we want to understand human consciousness or indeed animal consciousness overall, we can't just look to the brain. We need to look to the embodied, situated animal's life. No brain scan, no matter how cleverly constructed, is going to reveal the consciousness happening because that is not where the consciousness is happening.

That's the wrong level of analysis. The consciousness is un-folding in this dynamic. The consciousness is not in the head. There are a number of philosophers who are very sympathetic to this kind of extended conception of the mind. Andy Clark, for example, or Daniel Dennett—I view them very much as allies, al-though explicitly Andy doesn't think the extended mind approach for consciousness is valid. He thinks it works for cognition, certain kinds of cognitive processes, memory, and cognition, but not for consciousness.

The problem of consciousness is understanding how this world is there for us. It shows up in our senses. It shows up in our thoughts. Our feelings and interests and concerns are directed to and embrace this world around us. We think, we feel, the world shows up for us. To me that's the problem of consciousness. That is a real problem that needs to be studied, and it's a special problem.

A useful analogy is life. What is life? We can point to all sorts of chemical processes, metabolic processes, reproductive pro-cesses that are present where there is life. But we ask, where is the life? You don't say life is a thing inside the organism. The life is this process that the organism is participating in, a process that involves an environmental niche and dynamic selectivity. If you want to find the life, look to the dynamic of the animal's engage-ment with its world. The life is there. The life is not inside the animal. The life is the way the animal is in the world.

This is perhaps the biggest idea I can talk about with you today: the problem of consciousness and the problem of life are in effect the same problem, and the problem with so much of the science of consciousness today is that it treats consciousness as somehow separable from the mode of dynamic activity, which is the con-sciousness. (By the way, I should say this idea, this critical notion of the intimate interconnectedness of the problems of consciousness and life, is something that forms a theme of the work of Evan Thompson, who has a new book called *Mind in Life*.)

One way this comes out in an interesting way is if we look to a simple organism. An organism is not merely a collection of chemical processes. The organism has a certain unity, and it is only when I can conceptually bring that organism into focus as a unity that I can study it, that I can even recognize it. Once I do that, I can ask questions about what the organism's interests are, what its goals are, what its needs are. I can't ask about the needs of chemicals in a soup. There is a sense in which just to perceive the life in the thing before me, I need already to see it as an integrated whole distinct from its environment. Once I do that, I can also see it as having needs and interesting goals, and thus, in some sense, a mind. I don't mean to say that a bacterium has a mind. But I mean that wherever we find life we find the necessity for a certain kind of narrative which makes the attribution of mind at least intelligible.

This is the power of the theory of evolution: it makes the narrative official. Evolution shows us how life works; it allows us to tell stories about an organism that has the traits it has. We tell historical narratives. If we try to stay just at the level of atoms or molecules or chemical processes, we couldn't do that. So in a way my moral is this: the standpoint that cognitive science needs to take toward animals, and, indeed, toward ourselves, is the biological standpoint, the standpoint that allows us to bring the whole animal and its story into focus. Unfortunately, cognitive science has tended to take a distinctively nonbiological approach. They say they are looking at the brain and the nervous system, but they tend to model the brain and the nervous system as computational systems, systems thought of as solving problems and computing functions, systems that are, in the end, very much divorced from the active life of the animal.

Philosophers like to say that for all we know we could be a brain in a vat. But if you actually try to fill out the details of that thought experiment, it starts to seem much harder to make good sense of it. For example, very few of us would be inclined to think that a couple of cells in a petri dish were conscious. So how many more cells would we need to pile up before we began to think it became conscious? There is not any obvious way we can say where we would have to stop. It seems we would really need to try the experiment. But then who knows? It may be that we would have to build up to such a complex brain in a vat that what we ended up building is a brain and a virtual environment to house the brain. So maybe what this would teach us is that to make a mind you need to make a world. There would be consciousness in a world in a vat! Now, let's ask: where does the brain's body stop and the rest of the world begin? The critical point is that there's no way to draw this line a priori.

Evan Thompson and Diego Cosmelli have written a paper on this. They point out how much structure would need to go into the vat. The brain requires metabolism, it requires nourishment, and it requires the elimination of waste products. So if you actually try to fill in what the vat would look like, what you are actually describing is, in effect, a kind of body. But we already knew that a living brain and body can be conscious!

When we ask ourselves, wouldn't we have the same experience we are having now if we were being fed the right kind of stimulation, the answer is, yes, of course. But what does this show us? Again, we already know that there can be consciousness arising out of the close coupling of an animal and a world. But that's just what we are imagining when we imagine a mad scientist stimulating the brain. We are imagining a new kind of coupling of brain, body, and world. Crucially, here's what we are not describing: we are not describing a brain generating consciousness independently of the involvement of a world. We have not factored the world out of the equation. But that's what the old Cartesian thought experiment was aiming at, as if my internal states are sufficient for the world.

There is very interesting work done now in psychology labs— for example, work in O'Regan's lab—on the importance of eye movements and environmental stimulation for capturing attention and directing attention. This is why virtual reality systems are so hard to make really convincing. It's one thing to make a flight simulator—all you have to do is make a very good replica cockpit and a reactive virtual environment—but in most video games and in digital technology, there are huge shortcomings in the power of the virtual. In part this has to do with the fact our own perceptual attunement to the environment is so dependent on what the world brings to the table, as it were. Landmarks, markers, signposts that we respond to all play a role in our experiments. If you take the world out of the equation, I suspect that the brain, with its own internal powers, would be capable of producing only very impoverished experiences.

In fact, there is one nice bit of evidence I have to support that. There is a sleep scientist named Stephen LaBerge, who has done studies on lucid dreaming. In a series of studies that he did, he would interrogate these lucid dreamers on their experiences. I don't remember all the methodology, but the basic conclusion he found was this: in a dream it was impossible for one to look at a sign with text on it, look away, and look back at the sign, and have the sign say the same thing. In reality what enables us to look at a sign and look back and have it say the same thing is the reality. The sign anchors the experience. The sign carries the information. But the human brain on its own isn't good at storing information—if you look away, it's just impossible to see the same thing when you look back in a dream because in a dream you are responsible for all of that.

Our ordinary experience, the kind of richness, the texture, the stability of waking experience, can only be achieved for an organism that is actually locked into the environment in a certain kind of way. If you change the environment or take away the environment, you alter human consciousness. This points in a very profound way to this basic point I keep making, that the idea that you are your brain or that the brain alone is sufficient for consciousness is really just a mantra, and that there is no reason to believe it.

Alva Noë

14
Recursion and Human Thought: Why the Pirahã Don't Have Numbers

Daniel L. Everett

Former evangelical Christian missionary to the Pirahãs in the Brazilian Amazon for more than 20 years; linguistic researcher; Dean of Arts and Sciences, Bentley University; author, *Don't Sleep, There Are Snakes* and *Language*.

The research question that has motivated my work for the last 25 to 30 years has been, what is the nature of language? This is the question that motivates most linguistics research. But I started off asking it one way and came to the conclusion that asking it that way was probably wrong, and I now have a different way of approaching the problem.

My original concern was to think about Language with a capital "L." Human Language, what it's like in the brain, what the brain has to be like to sustain the capacity for Language. The most influential ideas for me in my early research were those of Noam Chomsky, principally the proposal that there is an innate capacity for grammar in our genes and that the acquisition of any given language is simply learning what the different parameter-settings are. What is a parameter?

Here's an example, called the "pro-drop" parameter. In English we always have to have a subject, even when it doesn't mean anything, like in "it rains"—"it" doesn't really refer to anything. "It" just is necessary because English has to have subjects. But in a language like Spanish or Portuguese, I don't say "it rains," I say just "rains"—*chuva*—in Portuguese, because Portuguese has a positive setting for the pro-drop parameter identified in Chomskyan research. All languages have either a positive (as Portuguese) or negative (as English) setting for this parameter. And it has other effects as well, in addition to allowing a language to drop subjects; it entails a number of other characteristics.

It's a very attractive idea that people are born with a genetic prespecification to set parameters in different ways, the environment serving as a "trigger." As Pinker put it, we have an instinct to learn language, and the environment triggers and shapes that instinct. But the environment is nothing more than that in this view—a shaper and a trigger; it is not fundamental to the actual final product in the Pinker-Chomsky view in the way that I have to come to think it actually is. Parameters and language as an instinct are very attractive ideas. Yet at the same time there are a number of components of languages that I've looked at that just don't seem to follow from these ideas.

The essence of human language is, according to Chomsky, the ability of finite brains to produce what he considers to be infinite grammars. By this he means not only that there is no upper limit on what we can say, but that there is no upper limit on the number of sentences our language has, there's no upper limit on the size of any particular sentence. Chomsky has claimed that the fundamental tool that underlies all of this creativity of human language is recursion: the ability for one phrase to reoccur inside another phrase of the same type. If I say "John's brother's house," I have a noun, "house," which occurs in a noun phrase, "brother's house," and that noun phrase occurs in another noun phrase, "John's brother's house." This makes a lot of sense, and it's an interesting property of human language.

Daniel L. Everett

But what if a language didn't show recursion? What would be the significance of that? First of all, it would mean that the language is not infinite—it would be a finite language, there could only be a limited number of sentences in that language. It would also mean that you could specify the upper size of a particular sentence in that language. That sounds bizarre, until we think of something like chess, which has also got a finite number of moves, but chess is an enormously productive game, it can be played and has been played for centuries, and many of these moves are novel, and the fact that it's finite really doesn't tell us much about its richness, or its importance.

If there were a finite language, because of the lack of recursion, that wouldn't mean that it wasn't spoken by normal humans, nor would it mean that it wasn't a very rich source of communication. But if you lived in an environment in which culture restricted the topics that you talked about, and not only just your general environmental limitations on the topics you talked about, but if there were a value in the culture that said, don't talk about topics that go beyond, say, immediate experience—in other words, don't talk about anything that you haven't seen or that hasn't been told to you by an eyewitness—this would severely limit what you could talk about. If that's the case, then that language might be finite, but it wouldn't be a poor language; it could be a very rich language. The fact that it's finite doesn't mean it's not a very rich language. And if that's the case, then you would look for evidence that this language lacked recursion.

So in the case of Pirahã, the language I've worked with the longest of the 24 languages I've worked with in the Amazon, for about 30 years, Pirahã doesn't have expressions like "John's brother's house." You can say "John's house," you can say "John's brother," but if you want to say "John's brother's house," you have to say

"John has a brother. This brother has a house." They have to say it in separate sentences.

As I look through the structure of the words and the structure of the sentences, it just becomes clear that they don't have recursion. If recursion is what Chomsky and Mark Hauser and Tecumseh Fitch have called "the essential property of language," the essential building block—in fact they've gone so far as to claim that that might be all there really is to human language that makes it different from other kinds of systems—then the fact that recursion is absent in a language—Pirahã—means that this language is fundamentally different from their predictions.

One answer that's been given when I claim that Pirahã lacks recursion is that recursion is a tool that's made available by the brain, but it doesn't have to be used. But then that's very difficult to reconcile with the idea that it's an essential property of human language—if it doesn't have to appear in a given language, then in principle, it doesn't have to appear in any language. If it doesn't have to appear in one part of a language, it doesn't have to appear in any part of a language.

It's not clear what causes recursion; in fact, just two weeks ago, at Illinois State University, we held an international conference on recursion in human language, which was the first conference of its kind ever held, and we had researchers from all around the world come and talk about recursion. One interesting thing that emerged from this is that the linguists, mathematicians, and computer scientists disagree on what recursion is, and how significant it is. Also, there are many examples of recursion lacking in a number of structures in languages where we otherwise would expect it. So recursion as the essential building block of human language, if Chomsky's correct, is difficult for me to apply as an intellectual trying to build a theory of human language, because it's not clear what it is, and it's not clear that it is in fact essential to different languages.

So as an alternative, what might we say? Well, recursion could occur because human beings are just smarter than species without it. In fact, the Nobel Prize–winning economist Herbert Simon, who taught psychology for many years at Carnegie Mellon University, wrote an important article in 1962 called "The Architecture of Complexity," and in effect, although he doesn't use this word, he argued that recursive structures are fundamental to information processing. He argued that these are just part of the human brain, and we use them not just in language but in economy, and discussion of problem solving, and the stories that we tell.

If you go back to the Pirahã language, and you look at the stories that they tell, you do find recursion. You find that ideas are built inside of other ideas, and one part of the story is subordinate to another part of the story. That's not part of the grammar per se, that's part of the way that they tell their stories. So my idea is that recursion is absolutely essential to the human brain, and it's a part of the fact that humans have larger brains than other species. In fact, one of the papers at the recursion conference was on recursion in other species, and it talked about how when deer look for food in the forest, they often use recursive strategies to map their way across the forest and back, and take little side paths that can be analyzed as recursive paths. So it's not clear, first of all, that recursion is unique to humans, and it's certainly not clear that recursion is part of language as opposed to part of the brain's general processing.

I am engaging in ongoing research on Pirahã, along with other researchers, including some from MIT's Brain and Cognitive Sciences department, led by Professor Ted Gibson, and other researchers from the University of Manchester. But my research is also part of a larger project funded by the European Commission, on characterizing human language by structural complexity, and the question we seek to answer there, with a number of researchers from Holland, Germany, and England, is, what is it that makes humans so smart, compared to other species? Is it just bigger brains? That might be the case. Or are there particular ways that our brain operates that makes it very different from the way that other kinds of brains operate?

Recursion has been proposed in human thinking to be the way that we think that other animals don't. That's very much an open empirical question, but let's say that it's right, in which case recursion once again underlies human thought, but doesn't have to make the jump into human language. You could in principle have a human language that is constrained by the culture, so that the language proper lacks recursion, but the brain has recursion. And that's very difficult to reconcile with Chomsky's ideas on where recursion comes from. Chomsky is absolutely correct to recognize the importance of recursion, but the role that he gives it, and the role that Hauser and Tecumseh Fitch give it, to me has got things backward. In other words, rather than going from language to the brain, we have to have recursion in language, and then it starts to make its manifestation in other thought processes. It starts in the thought processes and it might or might not jump to language. It does not seem to be an essential property of language, certainly not the essential property of language.

One prediction that this makes in Pirahã follows from the suggestions of people who worked on number theory and the nature of number in human speech: that counting systems—numerical systems—are based on recursion, and that this recursion follows from recursion in the language. This predicts in turn that if a language lacked recursion, then that language would also lack a

Daniel L. Everett

number system and a counting system. I've claimed for years that the Pirahã don't have numbers or accounting, and this has been verified in two recent sets of experiments, one of which was published in *Science* three years ago by Peter Gordon, arguing that the Pirahã don't count, and then a new set of experiments which was just carried out in January by people from Brain and Cognitive Sciences at MIT, which establishes pretty clearly that the Pirahã have no numbers, and, again, that they don't count at all.

So the evidence is still being collected, the claims that I have made about Pirahã lacking recursion and the fact that Pirahã is evidence that there probably isn't a need for universal grammar. Contrary to Chomsky's proposal that universal grammar is the best way to think about where language comes from, another possibility is just that humans have different brains that are different globally from those of other species, that they have a greater general intelligence that can be exploited for all sorts of purposes in human thinking and human problem solving. And one of the biggest problems we have to solve is how to communicate with other people—our conspecifics—and communication with our conspecifics is a problem that's often solved by recursion, but it doesn't have to be solved that way. It can be solved in other ways, especially in very small societies where so much information is implicit and held in common.

The ongoing investigation of these claims and alternatives to universal grammar, an architectonic effect of culture on grammar as a whole, and the implications of this for the way that we've thought about language for the last 50 years are serious. If I am correct, then the research so ably summarized in Steve Pinker's book *The Language Instinct* might not be the best way to think about things. Maybe there is no language instinct. So this is very controversial, and a lot more research has to be done. My colleagues and I are writing grants to test these claims. The only way that you can check out what I am saying is just to test the claims. Clearly formulate the claims and counterproposals, and go out and test them. If Everett's right, they ought to have this; and if he's wrong, we ought to find this. It's very simple conceptually to test the claims; you just have the logistical problems of the Amazon and a group that's monolingual and speaks no language but their own.

I don't think Pirahã is the only language that exhibits these qualities. What I think is that a lot of people are just like me in my beginning years of work there; they are given a set of categories to work with from their theories, and are told that these are the categories that languages have. So if you don't find a certain category, you just have to keep looking according to the theory. It takes a lot of courage, or, as in my case, frustration more than courage, to say, look, I'm not finding these things, so I'm just going to say they don't have the categories the theory predicts. Period. Say I am right about this. What are the implications?

I think that if we look at other groups, maybe groups in New Guinea and Australia, and some groups in Africa—what we have to find are groups that have been isolated, for various reasons, from larger cultures. The Pirahã's isolation is due to their very strong sense of superiority, and disdain for other cultures. Far from thinking of themselves as inferior because they lack counting, they consider their way of life the best possible way of life, and so they're not interested in assimilating other values.

They have another interesting value, which is "no coercion." That's one of the strongest Pirahã values: no coercion—you don't tell other people what to do.

I originally went to the Amazon to convert the Pirahã, to see them all become Christians, to translate the New Testament into their language. My only degree was an undergraduate degree from Moody Bible Institute in Chicago, and I went down there with the knowledge of New Testament Greek and a little bit of anthropology and linguistics.

When I first started working with the Pirahã, I realized that I needed more linguistics if I was going to understand their language. When I began to tell them the stories from the Bible, they didn't have much of an impact. I wondered, was I telling the story incorrectly? Finally one Pirahã asked me one day, well, what color is Jesus? How tall is he? When did he tell you these things? And I said, well, you know, I've never seen him, I don't know what color he was, I don't know how tall he was. Well, if you have never seen him, why are you telling us this?

I started thinking about what I had been doing all along, which was giving myself a social environment in which I could say things that I really didn't have any evidence for—assertions about religion and beliefs that I had in the Bible. And because I had this social environment that supported my being able to say these things, I never really got around to asking whether I knew what I was talking about. Whether there was any real empirical evidence for these claims.

The Pirahã, who in some ways are the ultimate empiricists—they need evidence for every claim you make—helped me realize that I hadn't been thinking very scientifically about my own beliefs. At the same time, I had started a PhD program in linguistics at the University of Campinas in southern Brazil, and I was now in the middle of a group of very intelligent Brazilian intellectuals, who were always astounded that someone at a university doing a PhD in linguistics could believe in the things I claimed to believe in at the time. So it was a big mixture of things involving the Pirahã, and at some point I realized that not only do I not have any evidence for these beliefs, but they have absolutely no applicability to these people, and my explanation of the universe.

I sat with a Pirahã once and he said, what does your god do? What does he do? And I said, well, he made the stars, and he made the Earth. And I asked, what do you say? He said, well, you know, nobody made these things, they just always were here. They have no concept of God. They have individual spirits, but they believe that they have seen these spirits, and they believe they see them regularly. In fact, when you look into it, these aren't sort of half-invisible spirits that they're seeing; they just take on the shape of things in the environment. They'll call a jaguar a spirit, or a tree a spirit, depending on the kinds of properties that it has. "Spirit" doesn't really mean for them what it means for us, and everything they say they have to evaluate empirically. This is what I hadn't been doing, and this challenged the faith that I thought I had, to the extent that I realized that it wasn't honest for me to continue to claim to believe these things when I realized how little investigation I had done into the nature of the things I claimed to believe.

I went to Brazil in 1977 as a missionary. I started my graduate program in 1979. By 1982, I was pretty sure that I didn't believe in the tenets of Christianity or any other religion or creeds based on the supernatural. But there's a social structure when you're a missionary, one that includes the income for you and your entire family as well as all of the relationships you've built up over the years. All the people you know and like and depend on are extremely religious and fundamentalist in their religion. It's very difficult to come out and say, "I don't believe this stuff any more." When I did say that, which was probably 13 years later, it had severe consequences for me personally. It's a difficult decision for anyone. I have a couple of friends whom I've told that it must be something like what it's like to come out as gay, to finally admit to your family that your values are just very very different from theirs.

My wife is still a missionary in Brazil to the Pirahã, and we've been separated for three years, and my view is pretty much irreconcilable with hers. It's difficult—it means that I don't go to that village when she's there. I don't go there and tell the Pirahãs not to believe in Jesus or anything like that. Actually I don't need to tell them that, because there's no danger that they ever will. They just find the entire concept—our beliefs—useless for them.

They wouldn't find the pope remotely impressive; they would find his clothes very impractical, and they would find it very funny. I took a Pirahã to Brasilia, the capital of Brazil, for health reasons once, to go to a hospital, and I took him to the Presidential Palace. As the president of Brazil was coming out, there was all this fanfare and I said, That's the chief of all Brazilians. Uh-huh. Can we go eat now? He was totally uninterested; the whole concept just sounds silly to them.

The first time I took a Pirahã on an airplane, I got a similar reaction. I was flying a man out for health reasons; he had a niece who needed surgery and he was accompanying her. We're flying above the clouds, and I know that he's never seen clouds from the top before, so I point down and I say, those are clouds down there. Uh-huh. He was completely uninterested; he acted like he flew in planes every day. The Pirahã are not that curious about what we have. They haven't shown interest in a number of things that other indigenous groups, even Amazonian groups, that have come out and had contact with in civilization for the first time are curious about. The Pirahã have been in regular contact for a couple of hundred years now, and they have assimilated almost nothing. It's very unusual.

The reason that I believe that the Pirahã are like this is because of the strong cultural values that they have—a series of cultural values. One principle is immediacy of experience; they aren't interested in things if they don't know the history behind them, if they haven't seen it done. But there's also just a strong conservative core to the culture; they don't change, and they don't change the environment around them much either. They don't make canoes. They live on the river, and they depend on canoes for their daily existence—someone's always fishing, someone's always crossing the river to hunt and gather—but they don't make canoes. If there are no Brazilian canoes, they'll take the bark off a tree and just sit in that and paddle across. And that's only good for one or two uses.

I brought in a Brazilian canoe master, and spent days with them and him in the jungle; we selected the wood, and made a dugout canoe. The Pirahã did all the labor—so they knew how to make a canoe, and I gave them the tools—but they came to me and they said, we need you to buy us another canoe. I said, well, we have the tools now, and you guys can make canoes. But they said, Pirahã don't make canoes. And that was the end of it. They never made a canoe like the Brazilians, even though I know that some of them have the skills to do that.

In the 1700s, the first Catholic mission to the Amazon area made contact with the Pirahã and the related people, the Muras, and abandoned them after a few years as the most recalcitrant group they had ever encountered. Other missionaries have worked with the Pirahã since then. Protestant missionaries have worked with them since about 1958, and there's not a single convert, there's not a single bit of interest.

A lot of people say that I'm a failure as a missionary. A lot of missionaries say I'm a failure—my ex-wife thinks I'm a failure as a missionary—and the reason they give is, I don't have enough faith. If you have enough faith, the story goes, God will overcome all of these things. But if you say that, you should know that god

Daniel L. Everett

is up against some serious cultural barriers. The Pirahã have a cultural taboo against talking about the world in certain ways, and the Christian message violates these.

They have the other cultural value against coercion that I mentioned. Religion is all about coercion—telling people how they should live and giving them a list of rules to live by—and the Pirahã just don't have coercion in that form. If someone were really violent and disrupted the entire life of the community, they would be ostracized; they might even be killed. But that would be a very serious pathological case in the culture. By and large, they tolerate differences, and even children aren't told what they have to do that much. Life is hard enough; if children don't do what they have to do, they'll go hungry. There's just no place for the Western concept of religion in their culture at all.

When a group receives this much publicity, you get different reactions. First of all, you get a lot of people who want to go there and investigate, until they see how difficult it is to get there, and how in fact they don't speak Portuguese and it's going to take a couple of years to be able to communicate with the Pirahã, even at a fairly simple level. This discourages people.

There are also a number of people who are upset that the group that they've been working with for 10 or 15 or 20 years didn't get any publicity.

Scientists—linguists and anthropologists in particular—are very reticent to say that one group is somehow more special than another group because if that's the case, then you've made discoveries that they haven't made. I really think that's probably right. I don't think the Pirahã are special in some deep sense. They're certainly very unusual, and they have characteristics that need to be explained, but all of the groups in the Amazon have different but equally interesting characteristics. I think that one reason we fail to notice, when we do field research, the fundamental differences between languages is that linguistic theory over the last 50 years—maybe even longer—has been primarily directed toward understanding how languages are alike, as opposed to how they are different.

If we look at the differences between languages—not exclusively, because what makes them alike is also very very important—the differences can be just as important as the similarities. We have no place in modern linguistic theory for really incorporating the differences and having interesting things to say about the differences. So when you say that this language lacks X, we will say, well, that's just an exotic fact: so they lack it, no big deal. But when you begin to accumulate differences across languages around the world, maybe some of these things that we thought were so unusual aren't as unusual and could in fact turn out to be similarities. Or the differences could be correlated with different components that we didn't expect before. Maybe there's something about the geography, or something about the culture, or something about other aspects of these people that account for these differences. Looking at differences doesn't mean that you throw your hands up and say there's no explanation and that you have nothing more than a catalog of what exists in the world. But it does develop a very different way of looking at culture and looking at language.

Missionaries have gone to the Pirahãs, learned their language more or less, and then left after a few years. There is a Brazilian anthropologist, Marco Antonio Gonçalves, who teaches at the University of Rio. He spent 18 months off and on working with them, and he speaks the language at a very basic level. The tones are part of what makes it so difficult for people who haven't had a lot of linguistic training, but it's just like Chinese, or Vietnamese, or Korean, in the fact that the tones are very important to the meanings of the words. This is really difficult for a lot of researchers without a significant linguistic background. I now have two researchers working with me from the University of Manchester: a PhD student who's writing her PhD on recursion or the absence thereof among the Pirahã; and a postdoc on a grant of mine whose research is looking at how well the Pirahã speak Portuguese, and if they do know some Portuguese, what kinds of grammatical characteristics does it have—does their Portuguese show anything that violates what I say about Pirahã itself? Both of these people, Jeanette Sakel and Eugénie Stapert, are learning Pirahã, and I've encouraged them to learn the language, and given them some lessons.

What happens with some people is that they go to the Pirahãs with me and I translate for them and help them get going. For people who do ongoing research of their own—many people have gone to the Pirahã with the idea that they're going to develop a multiyear research program and I'm going to be their partner every time they go.

I don't have the time to go with every researcher who wants to work with the Pirahã; I have my own research agenda. I've tried to help them to start learning the language, and most people sort of disappear after that. Tecumseh Fitch went with me last summer and he would like to go again, and maybe he will. But I think that the best way for anyone to go again is to invest the time to learn enough of the language to do their own research. There's also the fact that if anybody has to go with me, people can then say that my influence is so pervasive that you could never test what I'm saying, because I'm behind every single experiment.

Peter Gordon and I were colleagues at the University of Pittsburgh, and Peter did his PhD at MIT in psychology, with a strong concern for numerosity. We were talking, and I said, there's a group that doesn't count—I work with a group that doesn't count—and he found that very difficult to believe, so he wanted to go do experiments. He went, and I helped him get going; he did the experiments, but his explanation for the reason that the Pirahã don't count is that they don't have words for numbers. They only have one to many. I claim that in fact they don't have any numbers. His idea is that the absence of counting in Pirahã has a Whorfian explanation—that there's a linguistic determinism: if you lack numbers, you lack counting—that is, that the absence of the words causes the absence of the concepts. But this really doesn't explain a lot of things. There are a lot of groups that have been known not to have more than one to many—as soon as they got into a relationship where they needed it for trade, they borrowed the numbers from Portuguese or Spanish or English or whatever other language.

The crucial thing is that the Pirahã have not borrowed any numbers—and they want to learn to count. They asked me to give them classes in Brazilian numbers, so for eight months I spent an hour every night trying to teach them how to count. And it never got anywhere, except for a few of the children. Some of the children learned to do reasonably well, but as soon as anybody started to perform well, they were sent away from the classes. It was just a fun time to eat popcorn and watch me write things on the board. So I don't think that the fact that they lack numbers is attributable to the linguistic determinism associated with Benjamin Lee Whorf, i.e., that language determines our thought—I don't really think that goes very far. It also doesn't explain their lack of color words, the simplest kinship system that's ever been documented, the lack of recursion, and the lack of quantifiers, and all of these other properties. Gordon has no

Daniel L. Everett

explanation for the lack of these things, and he will just say, "I have no explanation, that's all a coincidence."

Some people have suggested that since this is a small society it's not unreasonable to hypothesize that there's a lot of inbreeding, and that this has made one particular gene much more prevalent in the society. Maybe Pirahã uniqueness is genetic in origin. People have asked me to do DNA tests, but my research has already been attacked for being borderline racist, because I say that the people are so different. So the last thing that I want to do is be associated with DNA testing. Somebody else can go there and do that. I don't think they have a closed gene pool, even though it's a small group of people. River traders come up frequently, and it's not uncommon for Pirahã to trade sex for different items off the boat that they want. So I don't think that genetics is relevant at all here.

Most inhabitants of the Brazilian Amazon are descended from Brazilian Indians, but now they would just consider themselves Brazilians. The Brazilians the Pirahãs most often see have boats, and they just come up the Pirahãs' river to buy Brazil nuts. In exchange, they bring machetes, gunpowder, powdered milk, sugar, whiskey, and so forth. The Pirahãs are usually interested in acquiring these things. They don't accumulate Western goods, but if you've got consumables, the Pirahã might buy, say, two pounds of sugar, pour it in a bowl, and eat it all at once. They're not going to put it on the shelf and save it; they'll just eat it when they get it.

It depends on the river trader, but sex is also a very common trade item. So you see these foreign babies being raised among the Pirahã. It's mainly the husband who works out the deal. Single women can negotiate on their own; wives wouldn't make that offer unless their husband negotiated it. In their dealings with outsiders, men take the lead, and the women won't usually come around unless they're called by Pirahã men. But promiscuity is not a problem for the Pirahã. It doesn't violate any values that they have.

I remember one time sitting in a hut with the Pirahã and they came and they said, we understand that you want to tell us about Jesus and that Jesus tells us that we should live in certain ways. Since you love Jesus, this is an American thing—but we don't want to live like you. We want to live like Pirahã, and we do lots of other things that you don't do, and we don't want to be like you. They've noticed these characteristics, and they much prefer to have the values that they have.

The paper I wrote that has attracted all the attention is "Cultural Constraints on Grammar and Cognition in Pirahã," published in the anthropology journal *Current Anthropology* in 2005. It would have been almost impossible to get this article—for one thing, it was 25,000 words, and for another thing it was so controversial—in a linguistics journal. Also, I chose this journal because it has a much higher circulation than any linguistics journal. And also the anthropology journal *Current Anthropology* invites commentators, a feature I really like. In my case, they had eight well-known linguists and anthropologists and psychologists comment on the paper. And the press picked up on it. You can never predict, obviously, when the press is going to pick up on something. But it started getting reported in magazines. And a lot of it was twisted; it wasn't exactly what I said, but it got a lot of play on radio, was in a lot of magazines. And everyone had the spin that this was—in fact I say it in the article—that this is a very strong counterexample to the kinds of claims that Chomsky makes. And I knew that there would be a response eventually, as it got more and more press.

David Pesetsky is a professor at MIT, Andrew Nevins was a student at MIT who now holds a temporary appointment in linguistics at Harvard, and Cilene Rodrigues is a Brazilian linguist

Daniel L. Everett

who I think is doing her PhD at MIT. They decided that they would write a reply to my article. The interesting thing is that I'm the main source of data on Pirahã. Now, the best way to check out what I'm saying would be to get some research funds to go down there and do experiments and test this stuff. But what they decided to do was to look at my doctoral dissertation, where I describe the grammar of Pirahã, and find inconsistencies between my doctoral dissertation and what I'm saying now. And there are some. And I say in the *Current Anthropology* article that there are inconsistencies, and that the 2005 article supersedes my previous work. And all of those decisions to change my mind on this or that analytical point were based on a lot of thought about what I had said previously and how it compared to my current knowledge.

My doctoral dissertation was written when I was using a certain set of grammatical categories common among most linguists, and I did my very best to make Pirahã come out and look like a "normal" language. So there are a couple of small examples of things that look like recursion in my doctoral dissertation. In fact I call them that. So the authors of the rebuttal dwell on these discrepancies. And then they try to counter my claims in the paper. Also, they refer to some unpublished studies by Steven Sheldon in which it is claimed that Pirahã has color words and number words. And they refer to an introduction to a dissertation on Pirahã that says that they speak Portuguese. And you do find these things in the studies they cite. But these are all written by people who either were not professional linguists, or who didn't speak the language. If you take the color words, Sheldon did in fact claim that the Pirahã had color words. But if you look at them, "mii sai," which he translated as "red," means "like blood." All of the color words in fact are just descriptions. This looks like blood, this looks like water, this looks like the sky, or this looks like a fire, or something like this. There can be any number of expressions. With regard to their ability to speak Portuguese, the Pirahã men do understand very simple Portuguese, just enough to trade with the river traders. Now if I went to Paris, I could probably get directions to the nearest bathroom, but that doesn't mean I speak French. I don't. That's roughly the Pirahãs' level of Portuguese.

So Pesetsky, Nevins, and Rodrigues were very careful in their criticisms, they worked very long and hard, they took months to do this. Then they posted it to a website called Ling Buzz, and it started being downloaded because of all the press on Pirahã—in the first few days there were 700 downloads. Every day it's getting dozens more downloads. I was actually trying to write something else at the time when I saw the reply, but I reluctantly put that aside to reply to their work. I replied to them point by point. The only part of their article that irritated me was the insinuations that because I focused on negative aspects of Pirahã, I was perhaps racist. They didn't use the term "racist," but they insinuated that I might have a negative view of the Pirahã as a people because I was only focusing on the gaps in the language. But I pointed out that I published more than 40 articles on Pirahã, and a book, and that all of those mainly talked about things they did have, not the gaps that they had. I put my reply on Ling Buzz, that's now the top-loaded paper on Ling Buzz, so those two papers are still getting downloaded a lot, and there's a debate going on. I don't know if they're planning a reply to my reply, but the way things go, they probably are.

When I saw this, I wrote to the three of them, and I said, you've put me in the interesting situation of pitting Dan Everett at 55 against Dan Everett at 26. Because, I said, all the data you use are my data. So I'll just have to explain why when I wrote my doctoral dissertation I didn't know as much about Pirahã as I know

now. Their objection is that even though I published extensively, I haven't published on all of these things previously. And so one of the many projects that I'm engaged in right now, along with several other people, mainly this group of researchers at MIT, Brain and Cognitive Sciences, is an experimental grammar of Pirahã, where we basically rewrite the grammar of Pirahã and do experiments to substantiate or test as many points as we can. If I had written all of this before I came out with the claims, I would never have come out with the claims. You have to make the claims and see the controversies, see what people say about them, to be sure you have the data. So I turned over all of my data to other researchers, and they're in the process of digitizing it, and eventually all these data will be on the Web, translated—it'll take a couple of years, but then you won't actually have to go to the Pirahã, you can look at the data, and you can search through the data and see if you can find counterexamples, or find other things that I've missed, and I'm sure people will.

When I was interviewed for *Der Spiegel*, I was at the Max Planck Institute for Evolutionary Anthropology in Leipzig, and my next-door neighbor was Tecumseh Fitch. We were talking about this quite a bit—he was my next-door neighbor both at the Institute and in the apartment building that we lived in, so we talked quite a bit, and went out a few times—and he made a comment about the fact that he didn't really believe the significance of what I was saying to the *Der Spiegel* reporter, so I wrote a reply to him, copying Noam Chomsky and Mark Hauser. And I actually thought that Tecumseh would be the one to respond, because that's who the letter was directed to, but in fact I immediately got a long response from Chomsky, followed by other long e-mails.

Initially it was a very interesting exchange. I know Noam fairly well, I've known his work for most of my career, and I've read everything he's ever written in linguistics—I could have written his responses myself. I don't mean to be flippant, but they were re-statements of things that everybody knows that he believes. I think that it's difficult for him to see that there is any alternative to what he's saying. He said to me that there is no alternative to universal grammar; it just means the biology of humans that underlies language. But that's not right, because there are a lot of people who believe that the biology of humans underlies language but that there is no specific language instinct. In fact at the Max Planck, Mike Tomasello has an entire research lab and one of the best primate zoos in the world, where he studies the evolution of communication and human language without believing in a language instinct or a universal grammar.

I've mainly followed Mike's research there because we talk more or less the same language, and he's more interested in di-rectly linguistic questions than just primatology, but there's a lot of really interesting work in primatology—looking at the acquisi-tion of communication and finding similarities that we might not have thought were there if we believed in a universal grammar.

I think that the way that Chomskyan theories developed over the last 50 years has made it completely untestable now. It's not clear what usefulness there is in the notion of universal grammar. It appeals to the public at large, and it used to appeal to linguists, but as you work more and more with it, there's no way to test it—I can't think of a single experiment. In fact I asked Noam this in an e-mail, what is a single prediction that universal grammar makes that I could falsify? How could I test it? What prediction does it make? And he said, it doesn't make any predictions; it's a field of study, like biology.

Now that is not quite right. No scientist can get by without be-lieving in biology, but it's quite possible to study human language without believing in universal grammar. So UG is really not a field of study in the same sense. I think the history of science shows that the people who develop a theory and who are responsible for the development of the theory are rarely the people who come forward and say: whoops, I was wrong, we need to actually work at it another way, this guy over here had the right idea. It's rare for that to happen. Noam is not likely to say this.

I want to have well-designed experiments to test my claims on recursion; I want to have mores studies of the Pirahã grammar from people working outside my influence. The more people who can look at this independently, the more likely it is that others are going to start to believe this, because I think it's going to be shown to be correct. If it's wrong, that's also important. The tests have to be done, and then if there is evidence that I might be onto something, we have to look at other languages, and other languages in similar situations where they've been cut off for one reason or another from outside influences for long periods of time. And re-examine those languages in light of the possibility that languages can vary more than we thought. And maybe the categories that we have aren't the best categories.

We need more fieldwork. Linguists have gotten away from fieldwork over the last 50 years. There's more interest in endangered languages now than there was a few years ago, but there's just now beginning to be a resurgence of the fieldwork ethic among linguists, and the idea that we can't figure out everything that we need to know just by looking at grammars that have been written, without going and seeing the language in the cultural context.

And that's really the biggest research question that I have for the future: what evidence is there that culture can exercise an architectonic effect on the grammar—that it can actually shape the very nature of grammar, and not simply trigger parameters?

15

The New Science of Morality:
An *Edge* Conference

Jonathan Haidt, Joshua Greene, Sam Harris, Roy Baumeister,

Paul Bloom, David Pizarro, Joshua Knobe

INTRODUCTION by John Brockman

Something radically new is in the air: new ways of understanding physical systems, new ways of thinking about thinking that call into question many of our basic assumptions. A realistic biology of the mind, advances in evolutionary biology, physics, information technology, genetics, neurobiology, psychology, engineering, the chemistry of materials: all are questions of critical importance with respect to what it means to be human. For the first time, we have the tools and the will to undertake the scientific study of human nature.

This began in the early seventies, when, as a graduate student at Harvard, evolutionary biologist Robert Trivers wrote five papers that set forth an agenda for a new field: the scientific study of human nature. In the past thirty-five years this work has spawned thousands of scientific experiments, new and important evidence, and exciting new ideas about who and what we are, presented in books by scientists such as Richard Dawkins, Daniel C. Dennett, Steven Pinker, and Edward O. Wilson, among many others.

In 1975, Wilson, a colleague of Trivers at Harvard, predicted that ethics would someday be taken out of the hands of philosophers and incorporated into the "new synthesis" of evolutionary and biological thinking. He was right.

Scientists engaged in the scientific study of human nature are gaining sway over the scientists and others in disciplines that rely on study-

ing social actions and human cultures independent from their biological foundation.

Nowhere is this more apparent than in the field of moral psychology. Using babies, psychopaths, chimpanzees, fMRI scanners, Web surveys, agent-based modeling, and ultimatum games, moral psychology has become a major convergence zone for research in the behavioral sciences.

So what do we have to say? Are we moving toward consensus on some points? What are the most pressing questions for the next five years? And what do we have to offer a world in which so many global and national crises are caused or exacerbated by moral failures and moral conflicts? It seems like everyone is studying morality these days, reaching findings that complement each other more often than they clash.

Culture is humankind's biological strategy, according to **Roy F. Baumeister**, and so human nature was shaped by an evolutionary process that selected in favor of traits conducive to this new, advanced kind of social life (culture). To him, therefore, studies of brain processes will augment rather than replace other approaches to studying human behavior, and he fears that the widespread neglect of the interpersonal dimension will compromise our understanding of human nature. Morality is ultimately a system of rules that enables groups of people to live together in reasonable harmony. Among other things, culture seeks to replace aggression with morals and laws as the primary means to solve the conflicts that inevitably arise in social life. Baumeister's work has explored such morally relevant topics as evil, self-control, choice, and free will. According to Yale psychologist **Paul Bloom**, humans are born with a hard-wired morality. A deep sense of good and evil is bred in the bone. His research shows that babies and toddlers can judge the goodness and badness of others' actions; they want to reward the good and punish the bad; they act to help those in distress; they feel guilt, shame, pride, and righteous anger.

Harvard cognitive neuroscientist and philosopher **Joshua D. Greene** sees our biggest social problems—war, terrorism, the destruction of the *environment, etc.—arising from our unwitting tendency to apply paleolithic moral thinking (also known as "common sense") to the complex problems of modern life. Our brains trick us into thinking that we have Moral Truth on our side when in fact we don't, and blind us to important truths that our brains were not designed to appreciate.*

University of Virginia psychologist **Jonathan Haidt**'s *research indicates that morality is a social construction which has evolved out of raw materials provided by five (or more) innate "psychological" foundations: Harm, Fairness, Ingroup, Authority, and Purity. Highly educated liberals generally rely upon and endorse only the first two foundations, whereas people who are more conservative, more religious, or of lower social class usually rely upon and endorse all five foundations.*

The failure of science to address questions of meaning, morality, and values, notes neuroscientist **Sam Harris**, *has become the primary justification for religious faith. In doubting our ability to address questions of meaning and morality through rational argument and scientific inquiry, we offer a mandate to religious dogmatism, superstition, and sectarian conflict. The greater the doubt, the greater the impetus to nurture divisive delusions.*

A lot of Yale experimental philosopher **Joshua Knobe**'s *recent research has been concerned with the impact of people's moral judgments on their intuitions about questions that might initially appear to be entirely independent of morality (questions about intention, causation, etc.). It has often been suggested that people's basic approach to thinking about such questions is best understood as being something like a scientific theory. He has offered a somewhat different view, according to which people's ordinary way of understanding the world is actually infused through and through with moral considerations. He is arguably most widely known for what has come to be called "the Knobe effect" or the "Side-Effect Effect."*

Disgust has been keeping Cornell psychologist **David Pizarro** *particularly busy, as it has been implicated by many as an emotion that plays* *a large role in many moral judgments. His lab results have shown that an increased tendency to experience disgust (as measured using the Disgust Sensitivity Scale, developed by Jon Haidt and colleagues) is related to political orientation.*

Jonathan Haidt

Social psychologist; Professor, New York University Stern School of Business; author, *The Happiness Hypothesis* and *The Righteous Mind*.

As the first speaker, I'd like to thank the Edge Foundation for bringing us all together, and bringing us all together in this beautiful place. I'm looking forward to having these conversations with all of you.

I was recently at a conference on moral development, and a prominent Kohlbergian moral psychologist stood up and said, "Moral psychology is dying." And I thought, well, maybe in your neighborhood property values are plummeting, but in the rest of the city, we are going through a renaissance. We are in a golden age.

My own neighborhood is the social psychology neighborhood, and it's gotten really, really fun, because all these really great ethnic groups are moving in next door. Within a few blocks, I can find cognitive neuroscientists and primatologists, developmental psychologists, experimental philosophers, and economists. We are in a golden age. We are living through the new synthesis in ethics that E. O. Wilson called for in 1975. We are living through an age of consilience.

We're sure to disagree on many points today, but I think that we here all agree on a number of things. We all agree that to understand morality, you've got to think about evolution and culture. You've got to know something about chimpanzees and bonobos and babies and psychopaths. You've got to know the differences between them. You've got to study the brain and the mind, and you've got to put it all together.

My hope for this conference is that we can note many of our points of agreement, as well as our disagreements. My hope is that the people who watch these talks on the Web will come away sharing our sense of enthusiasm and optimism, and mutual respect.

When I was a graduate student in Philadelphia, I had a really weird experience in a restaurant. I was walking on Chestnut Street, and I saw a restaurant called the True Taste. And I thought, well, okay, what is the true taste? So I went inside and looked at the menu. The menu had five sections. They were labeled "Brown Sugars," "Honeys," "Molasses" and "Artificials." And I thought this was really weird, and I went over to the waiter and I said, "What's going on? Don't you guys serve food?"

And it turns out, the waiter was actually the owner of the restaurant as well, and the only employee. And he explained to me that this was a tasting bar for sweeteners. It was the first of its kind in the world. And I could have sweeteners from 32 countries. He said that he had no background in the food industry, he'd never worked in a restaurant, but he was a PhD biologist who worked at the Monell Chemical Senses Center in Philadelphia.

And in his research he discovered that, of all the five taste receptors—you know, there's sweet, sour, salty, bitter, and savory—when people experience sweet taste, they get the biggest hit of dopamine. And that told him that sweetness is the true taste, the one that we most crave. And he thought, he reasoned, that it would be most efficient to have a restaurant that just focuses on that receptor, that will maximize the units of pleasure per calorie. So he opened the restaurant.

I asked him, "Well, okay, how's business going?" And he said, "Terrible. But at least I'm doing better than the chemist down the street, who opened a salt-tasting bar."

Now, of course, this didn't really happen to me, but it's a metaphor for how I feel when I read moral philosophy and some moral psychology. Morality is so rich and complex. It's so multifaceted and contradictory. But many authors reduce it to a single principle, which is usually some variant of welfare maximization. So that would be the sugar. Or sometimes, it's justice and related notions of fairness and rights. And that would be the chemist down the street. So basically, there's two restaurants to choose from. There's the utilitarian grille, and there's the deontological diner. That's pretty much it.

We need metaphors and analogies to think about difficult topics, such as morality. An analogy that Marc Hauser and John Mikhail have developed in recent years is that morality is like language. And I think it's a very, very good metaphor. It illuminates many aspects of morality. It's particularly good, I think, for sequences of actions that occur in time with varying aspects of intentionality.

But once we expand the moral domain beyond harm, I find that metaphors drawn from perception become more illuminating, more useful. I'm not trying to say that the language analogy is wrong or deficient. I'm just saying, let's think of another analogy, a perceptual analogy.

So if you think about vision, touch, and taste, for all three senses, our bodies are built with a small number of specialized receptors. So in the eye, we've got four kinds of cells in the retina to detect different frequencies of light. In our skin, we've got three kinds of receptors for temperature and pressure and tissue damage or pain. And on our tongues, we have these five kinds of taste receptor.

I think taste offers the closest, the richest, source domain for understanding morality. First, the links between taste, affect, and behavior are as clear as could be. Tastes are either good or bad. The good tastes, sweet and savory, and salt to some extent, these make us feel "I want more." They make us want to approach. They say, "This is good." Whereas sour and bitter tell us, "Whoa, pull back, stop."

Second, the taste metaphor fits with our intuitive morality so well that we often use it in our everyday moral language. We refer to acts as "tasteless," as "leaving a bad taste" in our mouths. We make disgust faces in response to certain violations.

Third, every culture constructs its own particular cuisine, its own way of pleasing those taste receptors. The taste analogy gets at what's universal—that is, the taste receptors of the moral mind—while it leaves plenty of room for cultural variation. Each culture comes up with its own particular way of pleasing these receptors, using local ingredients, drawing on historical traditions.

And fourth, the metaphor has an excellent pedigree. It was used 2,300 years ago in China by Mencius, who wrote, "Moral principles please our minds as beef and mutton and pork please our mouths." It was also a favorite of David Hume, but I'll come back to that.

So my goal in this talk is to develop the idea that moral psychology is like the psychology of taste in some important ways. Again, I'm not arguing against the language analogy. I'm just proposing that taste is also a very useful one. It helps show us morality in a different light. It brings us to some different conclusions.

As some of you know, I'm the codeveloper of a theory called Moral Foundations Theory, which specifies a small set of social receptors that are the beginnings of moral judgment. These are like the taste receptors of the moral mind. I'll mention this theory again near the end of my talk.

But before I come back to taste receptors and moral foundations, I want to talk about two giant warning flags—two articles published in *Behavioral and Brain Sciences*, under the wise editorship of Paul Bloom. And I think these articles are so important that the abstracts from these two articles should be posted in psychology departments all over the country, in just the way that, when you go to restaurants, they've got, you know, How to Help a Choking Victim. And by law, that's got to be in restaurants in some states.

So, the first article is called "The Weirdest People in the World," by Joe Henrich, Steve Heine, and Ara Norenzayan, and it was published last month in *BBS*. And the authors begin by noting that psychology as a discipline is an outlier in being the most American of all the scientific fields. Seventy percent of all citations in major psych journals refer to articles published by Americans. In chemistry, by contrast, the figure is just 37 percent. This is a serious problem, because psychology varies across cultures, and chemistry doesn't.

So, in the article, they start by reviewing all the studies they can find that contrast people in industrial societies with small-scale societies. And they show that industrialized people are different, even at some fairly low-level perceptual processing, spatial cognition. Industrialized societies think differently.

The next contrast is Western versus non-Western, within large-scale societies. And there, too, they find that Westerners are different from non-Westerners, in particular on some issues that are relevant for moral psychology, such as individualism and the sense of self.

Their third contrast is America versus the rest of the West. And there, too, Americans are the outliers, the most individualistic, the most analytical in their thinking styles.

And the final contrast is, within the United States, they compare highly educated Americans to those who are not. Same pattern.

All four comparisons point in the same direction, and lead them to the same conclusion, which I've put here on your handout. I'll just read it. "Behavioral scientists routinely publish broad claims about human psychology and behavior based on samples drawn entirely from Western, Educated, Industrialized, Rich, and Democratic societies." The acronym there being WEIRD. "Our findings suggest that members of WEIRD societies are among the least representative populations one could find for generalizing about humans. Overall, these empirical patterns suggest that we need to be less cavalier in addressing questions of human nature, on the basis of data drawn from this particularly thin and rather unusual slice of humanity."

As I read through the article, in terms of summarizing the content, in what way WEIRD people are different, my summary is this: the WEIRDer you are, the more you perceive a world full of separate objects rather than relationships, and the more you use an analytical thinking style, focusing on categories and laws, rather than a holistic style, focusing on patterns and contexts.

Now, let me state clearly that these empirical facts about "WEIRD-ness" don't in any way imply that our morality is wrong, only that it is unusual. Moral psychology is a descriptive enterprise, not a normative one. We have WEIRD chemistry. The chemistry produced by Western, Educated, Industrialized, Rich, Democratic societies is our chemistry, and it's a very good chemistry. And we have every reason to believe it's correct. And if a Ayurvedic practitioner from India were to come to a chemistry conference and say, "Good sirs and madams, your chemistry has ignored our Indian, you know, our 5,000-year-old chemistry," the chemists might laugh at them, if they were not particularly polite, and say, "Yeah, that's right. You know, we really don't care about your chemistry."

But suppose that same guy were to come to this conference and say, "You know, your moral psychology has ignored my morality, my moral psychology." Could we say the same thing? Could we just blow him off and say, "Yeah, we really don't care"? I don't think that we could do that. And what if the critique was made by an American evangelical Christian, or by an American conservative? Could we simply say, "We just don't care about your morality"? I don't think that we could.

Morality is like the Matrix, from the movie *The Matrix*. Morality is a consensual hallucination, and when you read the WEIRD people article, it's like taking the red pill. You see, oh my God, I am in one particular matrix. But there are lots and lots of other matrices out there.

We happen to live in a matrix that places extraordinary value on reason and logic. So, the question arises, is our faith justified? Maybe ours is right and the others are wrong. What if reasoning really is the royal road to truth? If so, then maybe the situation is like chemistry after all. Maybe WEIRD morality, with this emphasis on individual rights and welfare, maybe it's right, because we are the better reasoners. We had the Enlightenment. We are the heirs of the Enlightenment. Everyone else is sitting in darkness, giving credence to religion, superstition, and tradition. So maybe our matrix is the right one.

Well, let's turn to the second article. It's called "Why Do Humans Reason? Arguments for an Argumentative Theory," by Hugo Mercier and Dan Sperber. The article is a review of a puzzle that has bedeviled researchers in cognitive psychology and social cognition for a long time. The puzzle is, why are humans so amazingly bad at reasoning in some contexts, and so amazingly good in others?

For example, why can't people solve the Wason Four-Card Task, lots of basic syllogisms? Why do people sometimes do *worse* when you tell them to think about a problem or reason through it, than if you don't give them any special instructions?

Why is the confirmation bias, in particular—this is the most damaging one of all—so ineradicable? That is, why do people automatically search for evidence to support whatever they start off believing, and why is it impossible to train them to undo that? It's almost impossible. Nobody's found a way to teach critical thinking that gets people to automatically reflect on, well, what's wrong with my position?

And finally, why is reasoning so biased and motivated whenever self-interest or self-presentation are at stake? Wouldn't it be adaptive to know the truth in social situations, before you then try to manipulate?

The answer, according to Mercier and Sperber, is that reasoning was not designed to pursue the truth. Reasoning was designed by evolution to help us win arguments. That's why they call it the Argumentative Theory of Reasoning. So, as they put it, and it's here on your handout, "The evidence reviewed here shows not only that reasoning falls quite short of reliably delivering rational beliefs and rational decisions. It may even be, in a variety of cases, detrimental to rationality. Reasoning can lead to poor outcomes, not because humans are bad at it, but because they systematically strive for arguments that justify their beliefs or their actions. This explains the confirmation bias, motivated reasoning, and reason-based choice, among other things."

Now, the authors point out that we can and do reuse our reasoning abilities. We're sitting here at a conference. We're reasoning together. We can reuse our argumentative reasoning for other purposes. But even there, it shows the marks of its heritage. Even there, our thought processes tend toward confirmation of our own ideas. Science works very well as a social process, when we can come together and find flaws in each other's reasoning. We can't find the problems in our own reasoning very well. But that's what other people are for—to criticize us. And together, we hope the truth comes out.

But the private reasoning of any one scientist is often deeply flawed, because reasoning can be counted on to seek justification and not truth. The problem is especially serious in moral psychology, where we all care so deeply and personally about what is right and wrong, and where we are almost all politically liberal. I don't know of any conservatives. I do know of a couple of people in moral psychology who don't call themselves liberal. I think, Roy, are you one?

Okay. So there's you, and there's Phil Tetlock, who don't call themselves liberals, as far as I know. But I don't know anyone who calls themselves a conservative. We have a very, very biased field, which means we don't have the diversity to really be able to challenge each other's confirmation biases on a number of matters. So it's all up to you today, Roy.

So, as I said, morality is like the Matrix. It's a consensual hallucination. And if we only hang out with people who share our matrix, then we can be quite certain that, together, we will find a lot of evidence to support our matrix, and to condemn members of other matrices.

I think the Mercier and Sperber article offers strong empirical support for a basically Humean perspective— David Hume—a Humean perspective on moral reasoning. Hume famously wrote that "reason is and ought only to be the slave of the passions, and can never pretend to any other office than to serve and obey them." When Hume died, in 1776, he left us a strong foundation for what he and his contemporaries called "the moral sciences."

The subtitle of my talk today is "A Taste Analogy in Moral Psychology: Picking Up Where Hume Left Off." And at the bottom of the handout, I've listed some of the features that I think would characterize such a continuation, a continuation of Hume's project.

Hume was a paragon of Enlightenment thinking. He was a naturalist, which meant that he believed that morality was part of the natural world, and we can understand morality by studying human beings, not by studying Scripture or a priori logic. Let's look out at the world to do moral psychology, to do the moral sciences. So that's why I've listed naturalism, or naturalist, as the first of the seven features there.

Second, Hume was a nativist. Now, he didn't know about Darwin. He didn't know about evolution. But if he did, he would have embraced Darwin and evolution quite warmly. Hume believed that morals were like aesthetic perceptions, that they were "founded entirely on the particular fabric and constitution of the human species."

Third, Hume was a sentimentalist. That is, he thought that the key threads of this fabric were the many moral sentiments. And you can see his emphasis on sentiment in the second quotation that I have on your handout, where he uses the taste metaphor. He says, "Morality is nothing in the abstract nature of things, but is entirely relative to the sentiment or mental taste of each particular being, in the same manner as the distinctions of sweet and bitter, hot and cold arise from the particular feeling of each sense or organ. Moral perceptions, therefore, ought not to be classed with the operations of the understanding, but with the tastes or sentiments."

Now, some of these sentiments can be very subtle, and easily mistaken for products of reasoning, Hume said. And that's why I think, and I've argued, that the proper word for us today is not "sentiment" or "emotion." It's actually "intuition," a slightly broader term and a more sort of cognitive-sounding term.

Moral intuitions are rapid, automatic, and effortless. Since we've had the automaticity revolution in social psychology in the '90s, beginning with John Bargh and others, our thinking's turned a lot more toward automatic versus controlled processes, rather than emotion versus cognition. So intuition is clearly a type of cognition, and I think the crucial contrast for us in moral psychology is between various types of cognition, some of which are very affectively laden, others of which are less so, or not at all.

Fourth, Hume was a pluralist, because he was to some degree a virtue ethicist. Virtue ethics is the main alternative to deontology and utilitarianism in philosophy. Virtues are social skills. Virtues are character traits that a person needs in order to live a good, praiseworthy, or admirable life. The virtues of a rural farming culture are not the same as the virtues of an urban commercial or trading culture, nor should they be. So virtues are messy. Virtue theories are messy.

If you embrace virtue theory, you say good-bye to the dream of finding one principle, one foundation, on which you can rest all of morality. You become a pluralist, as I've listed down there. And you also become a nonparsimonist. That is, of course parsimony's always valuable in sciences, but my experience is that we've sort of elevated Occam's Razor into Occam's Chainsaw. Which is, if you can possibly cut it away and still have it stand, do it. And I think, especially in moral psychology, we've grossly disfigured our field by trying to get everything down to one if we possibly can. So I think, if you embrace virtue ethics, at least you put less of a value on parsimony than moral psychologists normally do.

But what you get in return for this messiness is, you get the payoff for being a naturalist. That is, you get a moral theory that fits with what we know about human nature elsewhere. So I often use the metaphor that the mind is like a rider on an elephant. The rider is conscious, controlled processes, such as reasoning. The elephant is the other 99 percent of what goes on in our minds, things that are unconscious and automatic.

Virtue theories are about training the elephant. Virtue theories are about cultivating habits, not just of behavior but of perception. So to develop the virtue of kindness, for example, is to have a keen sensitivity to the needs of other people, to feel compassion when warranted, and then to offer the right kind of help with a full heart.

Utilitarianism and deontology, by contrast, are not about the elephant at all. They are instruction manuals for riders. They say, "Here's how you do the calculation to figure out the right thing to do, and just do it." Even if it feels wrong. "Tell the truth, even if it's going to hurt your friends," say some deontologists. "Spend less time and money on your children, so that you have more time and money to devote to helping children in other countries and other continents, where you can do more good." These may be morally defensible and logically defensible positions, but they taste bad to most people. Most people don't like deontology or utilitarianism.

So why hasn't virtue ethics been the dominant approach? What happened to virtue ethics, which flourished in ancient Greece, in ancient China, and through the Middle Ages, and all the way up through David Hume and Ben Franklin? What happened to virtue ethics?

Jonathan Haidt

Well, if we were to write a history of moral philosophy, I think the next chapter would be called "Attack of the Systemizers." Most of you know that autism is a spectrum. It's not a discrete condition. And Simon Baron-Cohen tells us that we should think about it as two dimensions. There's systemizing and empathizing. Systemizing is the drive to analyze the variables in a system, and to derive the underlying rules that govern the behavior of a system. Empathizing is the drive to identify another person's emotions and thoughts, and to respond to these with appropriate emotion.

So, if you place these two dimensions, you make a 2x2 space, you get four quadrants. And autism and Asperger's are, let's call it, the bottom right corner of the bottom right quadrant. That is, very high on systemizing, very low on empathizing. People down there have sort of the odd behaviors and the mind-blindness that we know as autism or Asperger's.

The two major ethical systems that define Western philosophy were developed by men who either had Asperger's or were pretty darn close. For Jeremy Bentham, the principal founder of utilitarianism, the case is quite strong. According to an article titled "Asperger's Syndrome and the Eccentricity and Genius of Jeremy Bentham," published in the *Journal of Bentham Studies*, Bentham fit the criteria quite well. I'll just give a single account of his character from John Stuart Mill, who wrote, "In many of the most natural and strongest feelings of human nature, he had no sympathy. For many of its graver experiences, he was altogether cut off. And the faculty by which one mind understands a mind different from itself, and throws itself into the feelings of that other mind was denied him by his deficiency of imagination."

For Immanuel Kant, the case is not quite so clear. He also was a loner who loved routine, feared change, and focused on his few interests to the exclusion of all else. And according to one psychiatrist, Michael Fitzgerald, who diagnoses Asperger's in historical figures and shows how it contributed to their genius, Fitzgerald thinks that Kant would be diagnosed with Asperger's. I think the case is not nearly so clear. I think Kant did have better social skills, more ability to empathize. So I wouldn't say that Kant had Asperger's, but I think it's safe to say that he was about as high as he could possibly be on systemizing, while still being rather low on empathizing, although not the absolute zero that Bentham was.

Now, what I'm doing here, yes, it is a kind of an ad hominem argument. I'm not saying that their ethical theories are any less valid normatively because of these men's unusual mental makeup. That would be the wrong kind of ad hominem argument. But I do think that if we're doing history in particular, we're trying to understand, why did philosophy and then psychology—why did we make what I'm characterizing as a wrong turn? I think personality becomes relevant.

And I think what happened is that we had these two ultrasystemizers, in the late 18th and early 19th century. These two ultrasystemizers, during the early phases of the Industrial Revolution, when Western society was getting WEIRDer and we were in general shifting toward more systemized and more analytical thought. You had these two hypersystemized theories, and especially people in philosophy just went for it, for the next 200 years, it seems. All it is, you know, is utility, no? Deontology. You know, rights, harm.

And so you get this very narrow battle of two different systemized groups, and virtue ethics—which fit very well with the Enlightenment Project; you didn't need God for virtue ethics at all—virtue ethics should have survived quite well. But it kind of drops out. And I think personality factors are relevant.

Because philosophy went this way, into hypersystemizing, and because moral psychology in the 20th century followed them, referring to Kant and other moral philosophers, I think we ended up violating the two giant warning flags that I talked about, from these two *BBS* articles. We took WEIRD morality to be representative of human morality, and we've placed way too much emphasis on reasoning, treating it as though it was capable of independently seeking out moral truth.

I've been arguing for the last few years that we've got to expand our conception of the moral domain, that it includes multiple moral foundations, not just sugar and salt, and not just harm and fairness, but a lot more as well. So, with Craig Joseph and Jesse Graham and Brian Nosek, I've developed a theory called Moral Foundations Theory, which draws heavily on the anthropological insights of Richard Shweder.

Here I've just listed a very brief summary of it, that the five most important taste receptors of the moral mind are the following: care/harm, fairness/cheating, group loyalty and betrayal, authority and subversion, sanctity and degradation. And that moral systems are like cuisines that are constructed from local elements to please these receptors.

So I'm proposing, we're proposing, that these are the five best candidates for being the taste receptors of the moral mind. They're not the only five. There are a lot more. So much of our evolutionary heritage, of our perceptual abilities, of our language ability, so much goes into giving us moral concerns, the moral judgments that we have. But I think this is a good starting point. I think it's one that Hume would approve of. It uses the same metaphor that he used, the metaphor of taste.

In conclusion, I think we should pick up where Hume left off. We know an awful lot more than Hume did about psychology, evolution, and neuroscience. If Hume came back to us today, and we gave him a few years to read up on the literature and get up to speed, I think he would endorse all of these criteria. I've already talked about what it means to be a naturalist, a nativist, an intuitionist, a pluralist, and a nonparsimonist.

I just briefly want to say, I think it's also crucial, as long as you're going to be a nativist and say, "Oh, you know, evolution, it's innate," you also have to be a constructivist. I'm all in favor of reductionism, as long as it's paired with emergentism. You've got to be able to go down to the low level, but then also up to the level of institutions and cultural traditions and, you know, all kinds of local factors. A dictum of cultural psychology is that "culture and psyche make each other up." You know, we psychologists are specialists in the psyche. What are the gears turning in the mind? But those gears turn, and they evolved to turn, in various ecological and economic contexts. We've got to look at the two-way relations between psychology and the level above us, as well as the reductionist or neural level below us.

And then finally, the last line there. We've got to be very, very cautious about bias. I believe that morality has to be understood as a largely tribal phenomenon, at least in its origins. By its very nature, morality binds us into groups, in order to compete with other groups.

And as I said before, nearly all of us doing this work are secular liberals. And that means that we're at very high risk of misunderstanding those moralities that are not our own. If we were judges working on a case, we'd pretty much all have to recuse ourselves. But we're not going to do that, so we've got to just be extra careful to seek out critical views, to study moralities that aren't our own, to consider, to empathize, to think about them as possibly coherent systems of beliefs and values that could be related to coherent, and even humane, human ways of living and flourishing.

Jonathan Haidt

So that's my presentation. That's what I think the moral sciences should look like in the 21st century. Of course, I've created this presentation using my reasoning skills, and I know that my reasoning is designed only to help me find evidence to support this view. So I thank you for all the help you're about to give me in overcoming my confirmation bias, by pointing out all the contradictory evidence that I missed.

Joshua D. Greene

Cognitive neuroscientist; philosopher; Assistant Professor, Department of Psychology, Harvard University; author, *Moral Tribes*.

First, thanks so much to the Edge Foundation for bringing us together for this wonderful event. I want to talk about an issue that came up, quite conveniently, in a discussion of Jon Haidt's wonderful presentation. This is the issue of the "is" and the "ought," of descriptive moral psychology and normative prescription, a recommendation about how we should live.

Now, it's true that as scientists, our basic job is to describe the world as it is. But I don't think that that's the only thing that matters. In fact, I think the reason we're here, the reason we think this is such an exciting topic, is not that we think that the new moral psychology is going to cure cancer. Rather, we think that understanding this aspect of human nature is going to perhaps change the way we think and change the way we respond to important problems and issues in the real world. If all we were going to do is just describe how people think and never do anything with it, never use our knowledge to change the way we relate to our problems, then I don't think there would be much of a payoff. I think that applying our scientific knowledge to real problems is the payoff.

So, I come to this field both as a philosopher and as a scientist, and so my real core interest is in this relationship between the "is" of moral psychology, the "is" of science, and the "ought" of morality. What I'd like to do is present an alternative metaphor [to the one that Haidt offered], an alternative analogy, to the big picture of moral psychology.

Jon presented the idea of different taste receptors, corresponding to different moral interests. I think that it's not just about different tastes, or just about different intuitions. I think that there's a fundamental split between intuitions and reasoning, or between intuitions and more controlled processes. And I think Jon agrees with that.

From a descriptive point of view, I think Jon is absolutely right, and that he and others have done a wonderful job of opening our eyes to all of the morality that's out there that we've been missing. Descriptively, I think Jon couldn't be any more right.

But normatively, I think that there is something special about moral reasoning. And that while it may be WEIRD [in the sense outlined by Henrich and colleagues], it may be what we need moving forward, and that I think it's not just an accident, an obsession, that people, as the world started to become WEIRDer, started to take on people like Bentham and Kant as their leading lights. It makes sense that, as the world became WEIRDer, we started to need more reasoning, in a way that we didn't really need it before, when we were living in small communities. And that the problems that we face are going to require us to draw on parts of our psychology that we don't exercise very naturally, but that philosophers and scientists, I think, are at least more used to exercising, although not always in the service of progress.

Joshua D. Greene

So, the analogy: Like many of you, I have a camera. I'm not a real shutterbug, but I like my camera because it actually makes my life easier. My camera has a bunch of little automatic settings on it. That is, if you want to take a picture of someone in indoor lighting from about three feet away, you just put it in the portrait setting, and it configures all of the settings of the camera to do that pretty well. If you want to take a picture of a mountain in broad daylight from far away, you put it in the landscape setting, the action setting for sports, the night setting for night shooting. You get the idea. And it has these little preset configurations that work very well for the kinds of standard photographic situations that the manufacturer of the camera can anticipate.

But fortunately, the camera doesn't just have these point-and-shoot settings. It also has a manual mode. You can put it in manual mode, and you can adjust the F-stop and everything else yourself. And that's what you want to use if your goal, or your purposes, or your situation, is not the kind of thing that the manufacturer of the camera could anticipate. If you want to do something funky where you've got your subject off to the side, and you want the person out of focus, in low light, you have to put it in manual mode. You can't use one of the automatic settings to do something creative or different or funky.

It's a great strategy to have both automatic settings and a manual mode because they allow you to navigate the ubiquitous design tradeoff between *efficiency* and *flexibility*. The automatic settings give you efficiency. Point and shoot. And most of the time, it's going to work pretty well. The manual mode is not very efficient. You have to sit there and fiddle with it yourself. You have to know what you're doing. You can make mistakes. But in principle, you can do anything with it. It allows you to tackle new kinds of problems.

I think the human brain overall—not just when it comes to morality, but overall—uses the same design strategy as the camera that I described. We have automatic settings, which, as Jon Haidt suggested, you can think of as being like taste receptors that automatically make you say, "Ooh, I like that," or "Ooh, I don't like that."

But that's not all there is. We also have the ability to think consciously and deliberately and flexibly about new problems. That's the brain's manual mode. And I think that certain kinds of moral thinking and moral reasoning, while they do have roots in some of these taste receptors, they really owe much more to the elaboration of those taste receptors, through a kind of reasoning, through a kind of abstraction. And it's not just, "Pick your taste receptor."

I think, really, the biggest question is, are we going to rely on our intuitions, on our instincts, on our taste receptors? Or are we going to do something else? Now, some people might deny that there really is a "something else" that we can do. I disagree. I think that we can.

So, let me first give you a couple of examples of what I call "dual-process" thinking more generally. When I say dual-process, I mean the idea of having these automatic settings and having the manual mode.

And I should also say, with apologies to Liz and other people who emphasize this, this is an oversimplification. The brain is not clearly, simply divided into automatic and controlled. But I think that if you want to understand the world of the brain, this is the continent-level view. Before we can understand cities and towns and nations, we have to know what continent we're on. And I think those are the two big continents, the automatic and the controlled.

So, an example of automatic and controlled in the nonmoral domain comes from appetite and self-control. When it comes to

Joshua D. Greene

food, we all like things that are fatty and tasty and sweet, and a lot of us, at least, think that we might be inclined to eat too many of these things.

So, there's a nice study that was done by Baba Shiv, and Alexander—I hope I'm pronouncing this correctly—Fedurickin? [Fedorikhin] I'm probably not pronouncing it correctly. Very simple, elegant study. They had people come in for what they thought was a memory experiment. Some of them were told to remember a short little number. Some of them were told to remember a longer number. The longer number imposes what we call a "cognitive load." Keeps the manual mode busy.

And then they said, "Go down the hall to the other room, and by the way, there are snacks for you." And some of these snacks are yummy chocolate cake—they didn't tell them this up front—others are fruit salad. "Pick one snack and then go on your way." And what they predicted, and what they found, is that when people had the higher cognitive load, when the manual mode was kept busy, they were more likely to choose the chocolate cake, especially if they described themselves as being on a diet or looking to watch their weight.

And the idea is that we have an automatic tendency that says, "Hey! Yummy, fat, sweet things? Go for it!" And then another part of our brain that can say, "Well, that might be yummy, but in the long run, you'd rather be slim." And so there's this tension between the automatic impulse and the controlled impulse, which is trying to achieve some kind of larger goal for yourself.

I think we see the same kind of thing in moral psychology. And here I'll turn to what has recently become the sort of fruit fly of moral psychology. This is a classic example of going narrow and deep, rather than broad, but I think it's illuminating. And I know a lot of you are familiar with this, but for the uninitiated, I'll go through it. This is the Trolley Problem that philosophers have been arguing about for several decades now, and that in the last 10 years has become, as I've said, a kind of focal point for testing ideas in moral psychology.

So, the Trolley Problem, at least one version of it, goes like this: You've got a trolley that's headed toward these five people, and they're going to die if nothing is done. But you can hit a switch so that the trolley will turn away from the five and onto one person on another side track. And the question is, is it okay to turn the trolley away from the five and onto the one? And here, most people, about 90 percent of people, say that it is.

Next case, the trolley is headed toward five people once again. You're on a footbridge, over the tracks, in between the trolley and the five people, and the only way to save them, we will stipulate—somewhat unrealistic—is to push this large person—you can imagine, maybe a person wearing a giant backpack—off of the bridge and onto the tracks. He'll be crushed by the train, but using this person as a trolley-stopper, you can save the other five people. Here, most people say that this is not okay.

Now, there are a lot of things that are unrealistic about this case. It may not tell you everything you'd want to know about moral psychology. But there is a really interesting question here, which is, why do people quite reliably say that it's okay to trade one life for five in the first case, where you're turning the trolley away from the five and onto the one, but not okay to save five lives by pushing someone in front of the trolley, even if you assume that this is all going to work and that there are no sort of logistical problems with actually using someone as a trolley-stopper?

I and other people have looked at this almost every way possible now. A lot of different ways. With brain imaging, by looking at how patients with various kinds of brain damage respond to this,

Joshua D. Greene

with psychophysiology, with various kinds of behavioral manipulations.

And I think—not everyone here agrees with this—that the results from these studies clearly support this kind of dual-process view, where the idea is that there's an emotional response that makes you say, "No, no, no, don't push the guy off the footbridge." But then we have this manual mode kind of response that says, "Hey, you can save five lives by doing this. Doesn't this make more sense?" And in a case like the footbridge case, these two things conflict.

What's the evidence for this? As I said, there's a lot of different evidence. I'll just take what I think is probably the strongest piece, which is based on some work that Marc has done, and this has been replicated by other groups. If you look at patients who have emotion-related brain damage—that is, damage to a part of the brain called the ventromedial prefrontal cortex—they are four to five times more likely to say things like, "Sure, go ahead and push the guy off the footbridge."

And the idea is that if you don't have an emotional response that's making you say, "No, no, no, don't do this, this feels wrong," then instead, you're going to default to manual mode. You're going to say, "Well, five lives versus one. That sounds like a good deal." And that's, indeed, what these patients do.

Now, this is a short talk. I could go on at length. I usually go on for an hour, just talking about how trolley dilemmas and the research done in them supports this dual-process picture. If you want to ask me more about that, I'm happy to talk about it when we have a discussion, or later.

So this raises a more general question. Which aspects of our psychology should we trust, whether we're talking about the trolley problem or talking about some moral dilemma or problem in the real world? Should we be relying on our automatic settings, or should we be relying on manual mode?

And what I like about the camera analogy is that it points toward an answer to that question. The answer is not, "Automatic settings are good, manual mode is bad." The answer is not, "Manual mode is good, automatic settings are bad," which is what a lot of people think I think, but it's not what I think. They're good for different things.

Automatic settings are good for the kinds of situations where we have the right kind of training. And, I should say, where the instincts, the intuitions, have the right kind of training. Where they're the kinds of situations in which you can size it up efficiently and give an appropriate response. Where automatic settings are bad, or likely to be bad, is when we're dealing with a fundamentally new problem, one that we don't have the right kind of training for.

Okay. So, how do automatic settings get smart? How do they get the right kind of intelligence that's going to allow them to handle a situation well? There are three ways that this can happen.

First, an instinct can be what we think of paradigmatically as an instinct. That is, as something that's biologically entrained. So, if you have an animal that has an innate fear of its predators, that would be a case of learning happening on a biological, evolutionary scale. You have this instinct, as one of these creatures, because the other members of your species that didn't have it died. And so you have their indirect genetic inheritance. That's one way that an automatic response can be smart.

And that's not the only way. Another way is through cultural experience. Let me ask you, how many of you have had a run-in with the Ku Klux Klan, or with Nazis? I don't see any hands going up. But I bet that if I showed you pictures of swastikas, or men in pointy white hoods, a lot of you would have a flash of negative emotional response to these things. It doesn't come from biology. Biology doesn't know from Ku Klux Klan. And it doesn't come from individual learning. You've never had personal experiences with these groups. It comes from cultural learning. You can have instincts that are trained culturally, independent of your individual experience.

And finally, you could have individual experience, as when a child learns not to touch the hot stove. So biology, culture, and individual learning are all ways that our automatic settings can be trained up and can become adaptive.

Now, that covers a lot. But it doesn't cover everything. And I think there are two kinds of ways that we are now facing problems that our instincts, whether they're biologically, culturally, or individually trained, are not prepared for.

First of all, a lot of our problems, moral dilemmas, are the result of modern technology. For example, we have the ability to bomb people on the other side of the world. Or we have the ability to help people on the other side of the world. We have the ability to safely terminate the life of a fetus. We have the ability to do a lot of things that our ancestors were never able to do, and that our cultures may not have had a lot of trial-and-error experience with. How many cultures have had trial-and-error experience with saving the world from global warming? None. Because we're on trial number one, and we're not even through it yet. Right? So, that's one place.

Another place is intercultural contact. Our psychology, I think, is primarily designed for (a) getting along with people within our own group, and (b) dealing, either nicely or nastily, with members of other groups. And so the modern world is quite unusual, from an anthropological point of view, in terms of having people with different moral intuitions rubbing up against each other.

And so, as a result of these two main things, technology and cultural interchange, we've got problems that our automatic settings, I think, are very unlikely to be able to handle. And I think this is why we need manual mode. This is why, even if descriptively, careful, controlled moral reasoning has not been very important, if you took a catalog of all the moral thinking that's ever gone on, either recently or in human history, moral reasoning would be something like what Jon said, 1 percent. But moving forward, and dealing with the unique modern problems that we face, I think moral reasoning is likely to be very important.

And I think that we're too quick to use our point-and-shoot morality to deal with complicated problems that it wasn't designed, in any sense, to handle.

So, let me return for a moment. When do our intuitions do well and when do they not do well? First, let me go back to our fruit fly, the trolley problem. Once again, people say that it's okay to hit the switch. People say that it's not okay to push the guy off the footbridge. Why do we say that? What is it that we're responding to?

Well, one of the things that it seems like we're responding to is this: merely the difference between harming somebody in a physically direct way, versus not harming somebody in a physically direct way. So, if you give people the footbridge case, "Is it okay to push the guy off the footbridge?" at least in one version that I did, about 30 percent of people will say that it's okay. Not most people.

If you change it, ask a different group of people, "Suppose that you can drop this guy through a trapdoor in the footbridge, so that he'll just land on the tracks and get run over by the train, and that way you can save the five," the number doubles, to about 60 percent. You actually now get a narrow majority of people saying that it's okay. Now, this is not the only factor at work here, but it's

Joshua D. Greene

probably the single biggest factor. This factor is what I call "personal force"—the difference between pushing directly or doing something in a more mechanically mediated kind of way.

Now, we have this intuition. We don't even realize that this is what's affecting our judgment. But we can step back and reason and say, "Hmm . . . Is that why I said that there's a difference between these two cases? Does that really make sense?" You know, you might think, "Gosh, I wouldn't want to associate with someone who's willing to push somebody off of a footbridge," and that may be true. But consider this. Suppose someone called you from a footbridge and said, "Hey, here's this situation that's about to happen. Should I do this or not?" you would never say, "Well, it depends. Will you be using your hands, or will you be using a switch to land the guy on the track?" It's very hard to say that the presence or absence of "personal force" is something that matters morally, but it is something that our taste receptors are sensitive to, right? And so I think we can do better than our taste receptors, and it's not really just one taste receptor versus another. Let me give you some more real-world examples where I think this matters. About 30 years ago, Peter Singer posed to the philosophical world what is, I think, the most important moral problem we face. And he dramatized it with the following pair of cases:

You're walking by a shallow pond, and there's a child who's drowning there. And you could wade in and save this child easily, but if you do this, you're going to ruin your new Italian suit. (In something I wrote, I said, "It cost you $500." And my colleague Dan Gilbert says, "They cost a lot more than that, Josh." Two-thousand-dollar suit.) Now, you say, is it okay to let the child drown? Most of us would say, you're a moral monster if you say, "I'm going to let this child drown because I'm worried about my Armani suit."

Now, next case: There are children on the other side of the world who are desperately in need of food and medicine, and by making a donation smaller than $2,000, you can probably save at least one of their lives. And you say, "Well, I'd like to save them, but I have my eye on this Armani suit, and so I think I'm going to buy the Armani suit instead of saving them." There, we say, well, you ain't no saint, but we don't think that you're a moral monster for choosing to spend your money on luxury goods instead of saving other people's lives.

I think that this may be a case of emotional underreacting. And it makes sense from an evolutionary perspective. That is, we have emotional responses that are going to tug at our heartstrings when someone's right in front of us. But our heartstrings don't reach all the way to Africa. And so, it just may be a shortcoming of our cognitive design that we feel the pull of people who are in danger right in front of us, at least more than we otherwise would, but not people on the other side of the world.

Another example is what we're doing to the environment. If the environmental damage that we're doing—not just to the plants and atmosphere, but to our great-great-grandchildren, who we hope are going to live in the world—if that felt like an act of violence, we would probably be responding to our environmental problems very differently.

I think there are also cases where we're likely to be emotionally overreacting. Physician-assisted suicide, I think, is a nice example of this. And here, this is my opinion. This is not stating this as a scientific fact, in any of these cases. According to the American Medical Association, if somebody wants to die because they're in terrible pain and they have no good prospects for living it's okay to pull the plug. It's okay to allow them to die, to withhold or withdraw lifesaving support, but you're not allowed to give them something, even if they want it, that would actually kill them. Likewise, you can give them something—if it's morphine, let's say—to keep them comfortable, as long as your intention is to keep them comfortable, and you know that it's going to kill them. But your intention has to be to keep them comfortable.

I think that this is just a kind of moral squeamishness, and some cultures have come to this conclusion. It feels like a horrible act of violence to give somebody something that's going to kill them, and so, we say, this is a violation of the "sanctity of human life." I think this is a case where our emotions, or at least some people's emotions, overreact.

So, is there an alternative to point-and-shoot morality? I think that there is. And this is a much longer discussion, and I see I only have a few minutes left, so I'm only going to give you sort of a bare taste of this. But I think that what a lot of moral philosophy has done—and I think what we do as lay philosophers, making our cases out in the wider world—is we use our manual mode, we use our reasoning, to rationalize and justify our automatic settings.

And I think that, actually, this is the fundamental purpose of the concept of *rights*. That, when we talk about rights—"a fetus has a right to life," "a woman has a right to choose," Iran says they're not going to give up any of their "nuclear rights," Israel says, "We have a right to defend ourselves"—I think that rights are actually just a cognitive, manual-mode front for our automatic settings. And they have no real independent reality. This is obviously a controversial claim.

And I think the Kantian tradition [which gives primacy to rights] actually is manual mode. It's reasoning. But it's reasoning in the service of rationalizing and justifying those intuitions, as Jonathan Haidt has argued. Although Jon has argued, more broadly, that this is generally what goes on in moral discourse and moral philosophy.

I do think that there's a way out, and I think that it was people like Jeremy Bentham and, more recently, Peter Singer who've shown the way forward. Bentham is a geek. Bentham is, in many ways, emotionally and socially tone-deaf. But we have a real engineering problem here.

So, let me make an analogy with physics. If you want to get around the supermarket, you don't need an Einstein to show you how to navigate the aisles of the supermarket. Your intuitive physical intuitions will work pretty well. If you want to send a rocket to the moon, then you'd better put your physical instincts aside and do some geeky math.

And I think that what Bentham was doing was the geeky math of modern morality. Thinking, you know, by natural inclination, and in a way that's been very important to the modern world, thinking, "What here really matters? What can we really justify, and what seems to be just taste receptors that may or may not be firing in a way that, when we understand what's going on, will still make sense to us?"

And so, I'm sorry to leave you with such a vague prescription, but I'm out of time, and it's a big and complicated topic. But I think that geeky, manual-mode thinking is not to be underestimated. Because only a small part of the world may be WEIRD, in the Joe Henrich and Jon Haidt sense. But the world is getting WEIRDer and WEIRDer. We're dealing with cultures coming together with very different taste receptors, very different intuitions. We're dealing with moral problems that are created by modern technology, that we have no reliable, instinctive way of dealing with. The way I like to put it is that it would be a kind of *cognitive miracle* if our instincts were able to handle these problems.

And what I like about this idea is that—and this is what allows us to cross the is, ought divide—no matter what you think your standard for good or bad is, it's still a cognitive miracle if that standard is going to be built into those taste receptors. Because those taste receptors, whatever you think they ought to know, they couldn't possibly know it, because they don't have the biological, cultural, or individual experience to get things right in a point-and-shoot kind of way.

So I would say, in closing, that we shouldn't just be relying on our moral tastes. That there's a whole continent that may be not so important descriptively, but I think, moving forward, is going to be very important normatively. And it's important to understand it and to not discount it. And a better future may lie in a kind of geeky, detached, nonintuitive moral thinking that no one finds particularly comfortable but that we're all capable of doing, regardless of where we come from.

Sam Harris

Neuroscientist; Cofounder and CEO of Project Reason; author, *Letter to a Christian Nation*, *The Moral Landscape*, and *Free Will*.

What I intended to say today has been pushed around a little bit by what has already been said and by a couple of sidebar conversations. That is as it should be, no doubt. But if my remarks are less linear than you would hope, blame that—and the jet lag.

I think we should differentiate three projects that seem to me to be easily conflated, but which are distinct and independently worthy endeavors. The first project is to understand what people do in the name of "morality." We can look at the world, witnessing all of the diverse behaviors, rules, cultural artifacts, and morally salient emotions like empathy and disgust, and we can study how these things play out in human communities, both in our time and throughout history. We can examine all these phenomena in as nonjudgmental a way as possible and seek to understand them. We can understand them in evolutionary terms, and we can understand them in psychological and neurobiological terms, as they arise in the present. And we can call the resulting data and the entire effort a "science of morality." This would be a purely descriptive science of the sort that I hear Jonathan Haidt advocating.

For most scientists, this project seems to exhaust all the legitimate points of contact between science and morality—that is, between science and judgments of good and evil and right and wrong. But I think there are two other projects that we could concern ourselves with, which are arguably more important.

The second project would be to actually get clearer about what we mean, and should mean, by the term "morality," understand how it relates to human well-being altogether, and actually use this new discipline to think more intelligently about how to maximize human well-being. Of course, philosophers may think that this begs some of the important questions, and I'll get back to that. But I think this is a distinct project, and it's not purely descriptive. It's a normative project. The question is, how can we think about moral truth in the context of science?

The third project is a project of persuasion: how can we persuade all of the people who are committed to silly and harmful things in the name of "morality" to change their commitments, to have different goals in life, and to lead better lives? I think that this third project is actually the most important project facing humanity at this point in time. It subsumes everything else we could care about—from arresting climate change, to stopping nuclear proliferation, to curing cancer, to saving the whales. Any effort that requires that we collectively get our priorities straight and marshal massive commitments of time and resources would fall within the scope of this project. To build a viable global civilization we must begin to converge on the same economic, political, and environmental goals.

Obviously the project of moral persuasion is very difficult—but it strikes me as especially difficult if you can't figure out in what sense anyone could ever be right and wrong about questions of morality or about questions of human values. Understanding right and wrong in universal terms is Project Two, and that's what I'm focused on.

There are impediments to thinking about Project Two: the main one being that most right-thinking, well-educated, and well-intentioned people—certainly most scientists and public intellectuals, and, I would guess, most journalists—have been convinced that something in the last 200 years of intellectual progress has made it impossible to actually speak about "moral truth." Not because human experience is so difficult to study or the brain too complex, but because there is thought to be no intellectual basis from which to say that anyone is ever right or wrong about questions of good and evil.

My aim is to undermine this assumption, which is now the received opinion in science and philosophy. I think it is based on several fallacies and double standards and, frankly, on some bad philosophy. The first thing I should point out is that, apart from being untrue, this view has consequences.

In 1947, when the United Nations was attempting to formulate a universal declaration of human rights, the American Anthropological Association stepped forward and said, it can't be done. This would be to merely foist one provincial notion of human rights on the rest of humanity. Any notion of human rights is the product of culture, and declaring a universal conception of human rights is an intellectually illegitimate thing to do. This was the best our social sciences could do with the crematory of Auschwitz still smoking.

But, of course, it has long been obvious that we need to converge, as a global civilization, in our beliefs about how we should treat one another. For this, we need some universal conception of right and wrong. So in addition to just not being true, I think skepticism about moral truth actually has consequences that we really should worry about.

Definitions matter. And in science we are always in the business of framing conversations and making definitions. There is nothing about this process that condemns us to epistemological relativism or that nullifies truth claims. We define "physics" as, loosely speaking, our best effort to understand the behavior of matter and energy in the universe. The discipline is defined with respect to the goal of understanding how matter behaves.

Of course, anyone is free to define "physics" in some other way. A creationist physicist could come into the room and say, "Well, that's not my definition of physics. My physics is designed to match the Book of Genesis." But we are free to respond to such a person by saying, "You know, you really don't belong at this conference. That's not 'physics' as we are interested in it. You're using the word differently. You're not playing our language game." Such a gesture of exclusion is both legitimate and necessary. The fact that the discourse of physics is not sufficient to silence such a person, the fact that he cannot be brought into our conversation about physics, does not undermine physics as a domain of objective truth.

And yet, on the subject of morality, we seem to think that the possibility of differing opinions, the fact that someone can come forward and say that his morality has nothing to do with

Sam Harris

human flourishing—but depends upon following shariah law, for instance—the fact that such a position can be articulated proves, in some sense, that there's no such thing as moral truth. Morality, therefore, must be a human invention. The fact that it is possible to articulate a different position is considered a problem for the entire field. But this is a fallacy.

We have an intuitive physics, but much of our intuitive physics is wrong with respect to the goal of understanding how matter and energy behave in this universe. I am saying that we also have an intuitive morality, and much of our intuitive morality may be wrong with respect to the goal of maximizing human flourishing—and with reference to the facts that govern the well-being of conscious creatures, generally.

So I will argue, briefly, that the only sphere of legitimate moral concern is the well-being of conscious creatures. I'll say a few words in defense of this assertion, but I think the idea that it has to be defended is the product of several fallacies and double standards that we're not noticing. I don't know that I will have time to expose all of them, but I'll mention a few.

Thus far, I've introduced two things: the concept of consciousness and the concept of well-being. I am claiming that consciousness is the only context in which we can talk about morality and human values. Why is consciousness not an arbitrary starting point? Well, what's the alternative? Just imagine someone coming forward claiming to have some other source of value that has nothing to do with the actual or potential experience of conscious beings. Whatever this is, it must be something that cannot affect the experience of anything in the universe, in this life or in any other.

If you put this imagined source of value in a box, I think what you would have in that box would be—by definition—the least interesting thing in the universe. It would be—again, by definition—something that cannot be cared about. Any other source of value will have some relationship to the experience of conscious beings. So I don't think consciousness is an arbitrary starting point. When we're talking about right and wrong, and good and evil, and about outcomes that matter, we are necessarily talking about actual or potential changes in conscious experience.

I would further add that the concept of "well-being" captures everything we can care about in the moral sphere. The challenge is to have a definition of well-being that is truly open-ended and can absorb everything we care about. This is why I tend not to call myself a "consequentialist" or a "utilitarian," because traditionally, these positions have bounded the notion of consequences in such a way as to make them seem very brittle and exclusive of other concerns—producing a kind of body count calculus that only someone with Asperger's could adopt.

Consider the Trolley Problem: if there just is, in fact, a difference between pushing a person onto the tracks and flipping a switch—perhaps in terms of the emotional consequences of performing these actions—well, then this difference has to be taken into account. Or consider Peter Singer's Shallow Pond problem: we all know that it would take a very different kind of person to walk past a child drowning in a shallow pond, out of concern for getting one's suit wet, than it takes to ignore an appeal from UNICEF. It says much more about you if you can walk past that pond. If we were all this sort of person, there would be terrible ramifications as far as the eye can see. It seems to me, therefore, that the challenge is to get clear about what the actual consequences of an action are, about what changes in human experience are possible, and about which changes matter.

In thinking about a universal framework for morality, I now think in terms of what I call a "moral landscape." Perhaps there is a place in hell for anyone who would repurpose a cliché in this way, but the phrase "the moral landscape" actually captures what I'm after: I'm envisioning a space of peaks and valleys, where the peaks correspond to the heights of flourishing possible for any conscious system, and the valleys correspond to the deepest depths of misery.

To speak specifically of human beings for the moment: any change that can effect a change in human consciousness would lead to a translation across the moral landscape. So changes to our genome, and changes to our economic systems—and changes occurring on any level in between that can affect human well-being for good or for ill—would translate into movements within this hypothetical space of possible human experience.

A few interesting things drop out of this model: Clearly, it is possible, or even likely, that there are many peaks on the moral landscape. To speak specifically of human communities: perhaps there is a way to maximize human flourishing in which we follow Peter Singer as far as we can go, and somehow train ourselves to be truly dispassionate to friends and family, without weighting our children's welfare more than the welfare of other children, and perhaps there's another peak where we remain biased toward our own children, within certain limits, while correcting for this bias by creating a social system which is, in fact, fair. Perhaps there are a thousand different ways to tune the variable of selfishness versus altruism, to land us on a peak on the moral landscape.

However, there will be many more ways to not be on a peak. And it is clearly possible to be wrong about how to move from our present position to the nearest available peak. This follows directly from the observation that whatever conscious experiences are possible for us are a product of the way the universe is. Our conscious experience arises out of the laws of nature, the states of our brain, and our entanglement with the world. Therefore, there are right and wrong answers to the question of how to maximize human flourishing in any moment.

This becomes incredibly easy to see when we imagine there being only two people on earth: we can call them Adam and Eve. Ask yourself, are there right and wrong answers to the question of how Adam and Eve might maximize their well-being? Clearly there are. Wrong answer number one: they can smash each other in the face with a large rock. This will not be the best strategy to maximize their well-being.

Of course, there are zero-sum games they could play. And yes, they could be psychopaths who might utterly fail to collaborate. But, clearly, the best responses to their circumstance will not be zero-sum. The prospects of their flourishing and finding deeper and more durable sources of satisfaction will only be exposed by some form of cooperation. And all the worries that people normally bring to these discussions—like deontological principles or a Rawlsian concern about fairness—can be considered in the context of our asking how Adam and Eve can navigate the space of possible experiences so as to find a genuine peak of human flourishing, regardless of whether it is the only peak. Once again, multiple, equivalent, but incompatible peaks still allow for a realistic space in which there are right and wrong answers to moral questions.

One thing we must not get confused about is the difference between answers in practice and answers in principle. Needless to say, fully understanding the possible range of experiences available to Adam and Eve represents a fantastically complicated problem. And it gets more complicated when we add 6.7 billion to the experiment. But I would argue that it's not a different problem; it just gets more complicated.

Sam Harris

By analogy, consider economics. Is economics a science yet? Apparently not, judging from the last few years. Maybe economics will never get better than it is now. Perhaps we'll be surprised every decade or so by something terrible, and we'll be forced to concede that we're blinded by the complexity of our situation. But to say that it is difficult or impossible to answer certain problems in practice does not even slightly suggest that there are no right and wrong answers to these problems in principle.

The complexity of economics would never tempt us to say that there are no right and wrong ways to design economic systems, or to respond to financial crises. Nobody will ever say that it's a form of bigotry to criticize another country's response to a banking failure. Just imagine how terrifying it would be if the smartest people around all more or less agreed that we had to be nonjudgmental about everyone's view of economics and about every possible response to a global economic crisis.

And yet that is exactly where we stand as an intellectual community on the most important questions in human life. I don't think you have enjoyed the life of the mind until you have witnessed a philosopher or scientist talking about the "contextual legitimacy" of the burka, or of female genital excision, or any of these other barbaric practices that we know cause needless human misery. We have convinced ourselves that somehow science is by definition a value-free space, and that we can't make value judgments about beliefs and practices that needlessly derail our attempts to build happy and sane societies.

The truth is, science is not value-free. Good science is the product of our valuing evidence, logical consistency, parsimony, and other intellectual virtues. And if you don't value those things, you can't participate in the scientific conversation. I'm saying we need not worry about the people who don't value human flourishing, or who say they don't. We need not listen to people who come to the table saying, "You know, we want to cut the heads off adulterers at half-time at our soccer games because we have a book dictated by the creator of the universe that says we should." In response, we are free to say, "Well, you appear to be confused about everything. Your 'physics' isn't physics, and your 'morality' isn't morality." These are equivalent moves, intellectually speaking. They are borne of the same entanglement with real facts about the way the universe is. In terms of morality, our conversation can proceed with reference to facts about the changing experiences of conscious creatures. It seems to me to be just as legitimate, scientifically, to define "morality" in this way as it is to define "physics" in terms of the behavior of matter and energy. But most people engaged in the scientific study of morality don't seem to realize this.

Roy Baumeister

Francis Eppes Eminent Scholar and head of the social psychology graduate program, Florida State University; author, *Evil*.

John asked us to give some of our own personal quests or struggles or background. I don't know. The thing is, I was actually raised by wolves and it was not a happy childhood, you missed out on a lot of things. It taught me a lot but, you know, when I got to adolescence and I didn't want to be a wolf anymore, what they taught me was useless.

Ever since, I've been trying to figure out, so as not to miss out on any more, what human life was all about, and psychology is good for that. Figure out how the parts fit together. And to do that, one has to be something of a generalist, because I have to know what's in all the parts, and, well, today, to be a generalist, you know, you got to be fast because there's so little time, so much to know.

I go from area to area, trying to size things up. One thing I've learned is that caring about what the right answer is just slows you down. It gets in the way. And these are a lot of topics that people care very much about and have strong opinions about. I'd rather just not care. I aspire not to have political views.

Going from area to area, I notice some patterns that come up over and over again. One thing is that I've become increasingly skeptical of reductionism. It seems like reductionism is always proved wrong in the long run. In psychology we had behaviorism and Freudian psychoanalysis, which were going to explain everything. They explained some things and we learned a lot, but they could not explain everything. Far from it.

To some extent we're now going through this with the brain and evolution: many people think these will explain everything. Well, certainly we are going to learn a lot, and we have already. But we need to be attentive to continuities that we're the same as animals, and also perhaps the ways in which we are different, in order to put them together.

Beyond the reductionism, another thing is that motivation tends to be undervalued compared to cognition and ability. And a third point is that we tend to have the individual focus, so we tend to neglect and undervalue the interpersonal dimension, and things are perhaps more interpersonal than we are typically inclined to think.

And so that said, in terms of trying to understand human nature, and morality too, nature and culture certainly combine in some ways to do this, and I'd put these together in a slightly different way. It's not that nature's over here and culture's over there and they're both pulling us in different directions. Rather, nature made us for culture. I'm convinced that the distinctively human aspects of psychology, the human aspects of evolution, were adaptations to enable us to have this new and better kind of social life, namely culture.

Culture is our biological strategy. It's a new and better way of relating to each other, based on shared information and division of labor, interlocking roles and things like that. And it's worked. It's how we solve the problems of survival and reproduction, and it's worked pretty well for us in that regard. And so the distinctively human traits are often there to make this new kind of social life work.

Now, where does this leave us with morality? Well, it's not so much the purpose to facilitate individual salvation or perfection, or whatever, as I quoted McIntyre in our discussions earlier today, but rather, morality is the set of rules that enable people to live together. It serves the purpose of making the culture work, as culture depends on cooperating with each other—there's trust, shared assumptions, things like that.

Although nature and culture, in that sense, are working together, there are some conflicts; in particular, nature's made us, at least in a very basic way, selfish. The brain is selfish, and maybe it's the selfish gene, not the selfish individual, or whatever. But there's still a natural selfishness, whereas culture needs people to overcome this to some degree, because you have to cooperate with others and do things that are detrimental to your short-term, and even your long-term, self-interest. In order for culture to work, you have to keep your promises, you have to wait your turn, pay your taxes, maybe even send your offspring into battle to risk their lives. It goes against the grain, biologically. But these are the sorts of things that morality promotes, to try to get people to overcome their natural selfish impulses, to do things that make the system work. And that benefits everyone in the long run.

Morality does this, and of course laws, too. We haven't said that much about laws, but laws regulate behavior in a lot of the same ways that morality does. They prescribe a lot of the same things, restraining self-interest to do what is better for the group so that the system will operate effectively. And there's a big difference between laws and morals, which is mainly in the force they use. Why people have to do moral things in practice is because of concern with their reputation, and it's based, therefore, on long-term relationships. If you cheat someone you're living next door to, for the rest of your life they're going to know that, and other people are going to know that, and you'll be punished, and it will compromise your outcomes long-term.

As society got larger and more complex and moved to more interactions with strangers, laws have had to step in to take their place, because you can cheat a stranger whom you'll never see again and get away with it. Anyway, you're seeing here the neglected interpersonal dimension in understanding morality. Morality depends on relationships. And it's there, again, to regulate interpersonal behavior so that people cooperate, so that the system can work.

Now, consider some of the traits that evolved to enable people to overcome these selfish impulses so as to do what's best for the group and the system and so on. Among those, self-regulation is central. I think in part I got invited here because I have a history of doing research on self-regulation and self-control. The essence of self-regulation is to override one response so that you can do something else—usually something that's more desirable, better either in the long run or better for the group.

That is why we've called self-control the moral muscle. I'm going to unpack that and comment on both parts. It's moral:

self-control is moral in the sense that it enables you to do these morally good things, sometimes detrimental to self-interest. So if you get lists of morals, whether it's the seven deadly sins or the Ten Commandments or a list of virtues and so on, they're mostly about self-control. And you can really see self-control as central to them, so there are the seven deadly sins of gluttony, wrath, greed, and the rest. They're mostly self-control failures. Likewise, the virtues are exemplary patterns of self-control. So that's the moral part of the moral muscle—it's a capacity to enable us to do these moral actions, which are good for the group, even though overcoming this short-term self-interest.

The muscle part—that's kind of emerged from our lab work, independent of any moral aspect. There seems to be a limited capacity to exert self-control that gets used up. It's like a muscle, it gets tired. As we found in many studies, after people do some kind of self-control task, then go to a different context with completely different self-control demands, they do worse on it—as if they used a muscle and it got tired.

So it's a limited resource that gets exhausted. There are other aspects of the muscle analogy. If you exercise self-control regularly, you get stronger. I wouldn't want people to say, well, if self-control and morality's a limited capacity, I'm never going to do anything to exert self-control because I don't want to waste it. No, au contraire, you should exert it regularly; it will make you stronger and give you greater capacity to do things.

And certainly then we find that when people have exerted this muscle and it's tired, so to speak, or when they've depleted—you know, ego depletion's a term for it—depleted their resources, then behavior drifts toward being less moral. So we found that people are perhaps more gratuitously aggressive toward somebody else

Roy Baumeister

after they've exerted self-control and used up some of their moral muscle resources.

In a study on cheating and stealing we published a couple of years ago, people had to type up an essay about what they had done recently, either not using words with the letter "a" or not using words containing the letter "x." There are a lot more words that contain an "a" than "x," so the former requires much more self-control and overriding. And so when you're trying to make up a sentence and you keep reaching the point, oh look, there's an "a" in that word and you have to override it, and it uses self-control to keep overriding one response and coming up with another, that depletes people's resources. So they were more depleted in the "a" than in the "x" condition.

Afterward, they went to another room, supposedly another experiment where they're taking an arithmetic test and they're being paid for the number of ones they get right. They either scored it themselves, or the experimenter scored it for them. Of the four conditions (depleted or not, and self-scored or experimenter-scored) all got about the same number right—except for the depleted people who scored their own tests. They somehow claimed to get a whole lot more right. It was not plausible that they were actually getting smarter by virtue of having typed while not using words with the letter "a" in them, because when the experimenter scored them, he couldn't find any difference. Got about the same number right. But when nobody was checking and their answer sheet was shredded and they said, you know, I got six correct, then suddenly they got a whole lot more correct. So that suggests an increase in lying and cheating, and effectively stealing money from the experiment.

There are some other findings, too: depleted people are more likely to engage in sexual misbehavior, and so on. So moral behavior does seem to go down when people have depleted their moral muscle capacity. More recently, we're working with Marc Hauser on seeing if depletion changes how people make moral judgments of others—that's proving a little bit more slippery. But again, this kind of process is geared toward regulating your behavior more than your thinking about others. So it's not surprising that it shows up right there.

A couple of other things we've found are relevant here. Choice seems to deplete the same muscle as self-control; it's the same resource. So we have people make a lot of choices about which of these two products would you buy and so on, and afterward their self-control is damaged, too, so making choices uses up the resource needed for self-control. That resource seems to be tied into some physiological processes. We found changes with the glucose levels in the bloodstream, and so something about doing these advanced kinds of self-control acts uses up this resource and depletes self-control in the bloodstream.

If you give people a drink—after manipulation we give them lemonade mixed with sugar or with Splenda, and with Splenda they still act bad. But if they get sugar in there, it gives a quick dose of glucose to the bloodstream and suddenly their behavior is more self-controlled, in some cases more moral—making more rational decisions and so forth. And conversely, too, if they're depleted from self-control, then their choice process is changed to be more shallow and so forth.

In terms of self-regulation plus choice, I mean, you start now to think that this same capacity is used for choosing and for self-control, and in maybe a couple of other things as well. There are some data on initiative. So instead of talking about it in terms of regulatory depletion, we're trying to come up with a bigger term, and that's how I got to talking about free will.

Free will is another of these topics where people are very emotional on both sides, and they have a lot of passionate feelings. And I don't really want to deal with that. Let me try to forestall some by saying that when I'm trying to develop a scientific theory of free will, there's nothing supernatural, nothing that's noncausal in there. Let's understand the processes by which people make these choices and exert self-control. And I think there is a social reality corresponding to this, and certainly behaving with self-control, behaving morally, making moral choices, and making certain kinds of choices are the things we associate with free will. And so in that sense, there's a real phenomenon there. Whether it deserves the term "free will" depends kind of on this or that definition.

I'm surprised—I've been to this conference, I was at this conference in Israel with Bloom and Pizarro on morality, and yet nobody at either conference mentioned free will, really, in any talk or discussion. Yet it seems to me that this is a natural way to build this theory and extend it. So part of my interest in this topic—morality assumes that the person can do different things. And it says, well, this act is good and that act is bad, so it's a way to persuade you to do one thing rather than the other.

And likewise, moral judgments about people are based on the assumption that the person could have, and essentially the moral judgment says the person should have, acted differently. And legal judgments, of course, are very much the same sort of thing.

So I see I'm well ahead of schedule here. Let me comment on a couple of other points. In terms of evolution and morality, there was a recent article by David Barash saying, well, there's the fairness instinct, you can see it in other animals; and he cited Frans De Waal's study, in which monkeys were mad if they saw another monkey getting a nicer treat for the same action. They'd say, well, look, I think there's a fairness instinct. Again, I'm skeptical of reductionism, and you know, we need to attend to both the continuities and also the differences between human and animal behavior. To call it a fairness instinct seems a little overstated; it's a step in that direction, but it's not that impressive.

If you have two dogs and you give one of them a treat, the other looks at you like, well, what about me? But what you don't see is the overbenefited one complain. The other dog doesn't say, well, I'll share my biscuit with you, or I'm not going to eat mine until the other dog gets one too. Yet human behavior does show some of those patterns. And so I think if we want to see a fairness instinct, we need to see both the overbenefited and the underbenefited one complain. And perhaps even more: to get to the human, you have to have a third party saying no, you got more than this one and that's not fair, and intervening to redistribute, as happens all over the world in human societies.

In the Israel conference, Paul Bloom was talking about moral progress, too. And Steven Pinker has a recent book on that as well, I gather. Yes, the world's gotten to be a better place, but again, I'm not sure that we're morally better people. The laws are very much responsible for accomplishing that. It's a lot of third-party intervention to tell people not to do that. That reflects really some things that are new in human culture, perhaps not seen so much in other creatures.

Let me draw some conclusions here. Culture, I want to say, is humankind's biological strategy. It's our new way of solving the basic biological problems of survival and reproduction. We take our sick children to the hospital; we ask the government to give tax breaks for research or to provide tax breaks to families with children or whatever. It's been very successful, culture. It worked very well for us, but requires a lot of advanced psychological traits.

One might ask, if culture works so well, why don't other species use it? Well, they don't have as many capacities. Culture requires advanced psychological capabilities. And so human evolution maybe added some new things, or at least took what was small in other animals and made it larger and more central. Self-control is present in other animals, but needs to be developed much more thoroughly in humans because culture has a lot more rules, a lot more regulations, of the laws and morals and so on. So a lot more needing to override your behavior to bring it into line with standards.

Morality in the full-fledged sense, and I'm going with the cultural materialist view that culture is a system that basically has to provide for the material and social needs of the individuals. And so regulates behavior for that, and morality comes with it, in a full-fledged sense, comes with culture. Tells people what to do to override their self-interest, and at least their short-term, and to follow the system's rules. The system works, and because of that we all live better, but we all have to cooperate to a significant degree in order for the system to work. And so morality is this set of rules to help us do that.

Self-control, then, is one of the crucial mechanisms that had to improve in humans, to enable culture to succeed. So it's an inner capacity, limited energy expense, and so on, to alter your behavior, override responses, and enable one to change one's behavior to fit in with the requirements of the system so that it will work. And then free will, again, you can see continuity with animals, their choice and agency in other creatures, and free will perhaps a more advanced form of agency that evolved out of that, and more adapted to working in culture using meaningful reasons and operating within the context of the shared group.

It enables the human animal to relate to its social and cultural environment. I mean, a simple way of the basic agency of the squirrel and so on is to enable that little animal to deal with its physical environment, but the free will as an advanced form enables the human being to deal with its cultural environment. And recognizes that as humans we can be somewhat more than animals, control our behavior in these advanced ways, need to make the system work. And once it works for us, then it has provided the immense benefits that it has.

Paul Bloom

Brooks and Suzanne Ragen Professor of Psychology, Yale University; author, *Descartes' Baby*, *How Pleasure Works*, and *Just Babies*.

I'd like to thank the Edge Foundation for putting together this workshop, and I'd also like to thank all of my colleagues here. It's because of the extraordinary theoretical and empirical work of the people in this room that the study of morality is, I think, the most exciting field in all of psychology. So I'm really glad to be included among this group.

What I want to do today is talk about some ideas I've been exploring concerning the origin of human kindness. And I'll begin with a story that Sarah Hrdy tells at the beginning of her excellent new book, *Mothers and Others*. She describes herself flying on an airplane. It's a crowded airplane, and she's flying coach. She waits in line to get to her seat; later in the flight, food is going around, but she's not the first person to be served; other people are getting their meals ahead of her. And there's a crying baby. The mother's soothing the baby, the person next to them is trying to hide his annoyance, other people are coo-cooing the baby, and so on.

As Hrdy points out, this is entirely unexceptional. Billions of people fly each year, and this is how most flights are. But she then imagines what would happen if every individual on the plane was transformed into a chimp. Chaos would reign. By the time the plane landed, there'd be body parts all over the aisles, and the baby would be lucky to make it out alive.

The point here is that people are nicer than chimps. Human niceness shows up in all sorts of other ways. Americans give hundreds of billions of dollars each year to charity. Now, you might be cynical about some of that giving, but some of it seems to be genuinely motivated by concern for strangers. We leave tips at restaurants. We leave tips in our hotel rooms. This last one is striking: some of us, when leaving this hotel, will leave money for the maid, even though this act has no possible selfish benefit. It doesn't help our reputation; it won't improve future service. We do it anyway, because we feel that it is right.

My favorite experiment on adult human niceness was done by Stanley Milgram many years ago. Milgram was a Yale psychologist who is most famous for his obedience experiments, where he found that people would kill strangers if asked to do so in the right way. But he was also interested in niceness, and he did an experiment in which he left stamped envelopes scattered around New Haven. The question was how many of them would be delivered. And the answer was well over half. Now, it wasn't indiscriminate: if instead of a person's name on the letter, it was "Friends of the Nazi Party," people wouldn't deliver it. Presumably they'd look at it, they'd throw it in the garbage, they'd say to hell with that.

In a more recent study, another psychologist replicated the study but didn't even put stamps on the letters. Still, one in five letters came back. This is extraordinarily nice.

I'm a developmental psychologist and I'm interested in where this niceness comes from. It turns out that at least some of it seems to be hard-wired, emerging naturally. It is not taught.

The idea here was anticipated by Adam Smith hundreds of years ago. Adam Smith was the founder of modern economics, and he was very sophisticated when it came to human sentiment. He pointed out that when you see somebody in pain, you feel their pain—to at least some extent—as if it was yours. And you're motivated to make it go away, you're motivated to help. This is a primitive good that doesn't reduce to any other good.

It turns out that some such empathy exists even in babies. When babies hear crying, they'll start to cry themselves. Now, some very cynical psychologists worried that this isn't empathy at all. It's because babies are so stupid that when they hear another baby crying, they think they're crying themselves, so they get upset and they cry some more. In response, though, other psychologists did experiments where they exposed babies to tape-recorded sounds of their own cries and tape-recorded sounds of other babies' cries. And they found that the babies cry more to the sounds of other babies than to their own cries, suggesting that this response really is other-directed. Furthermore, when a baby sees someone in pain, even silent pain, the baby will get distressed. And as soon as babies are old enough to move around their bodies, they'll try to make the pain go away. They'll stroke the other person, or they'll try to hand over a toy or a bottle.

In some recent work that Roy Baumeister mentioned in passing, Felix Warneken and Michael Tomasello set up a clever experiment where they put toddlers in situations where nobody is looking at them, and then an adult comes in and has some sort of minor crisis, such as reaching for something and being unable to get to it, or trying to get access to a cabinet with his arms too

Paul Bloom

full to open the door. And Warneken and Tomasello find that toddlers, more often than not, will spontaneously toddle over and try to help.

In my own research, I've been interested not so much in moral action or altruistic behavior, but in moral cognition, moral intelligence. And this is a series of studies that I've been doing in collaboration with Karen Wynn, my colleague at Yale, who runs the Yale Infant Lab, and a wonderful graduate student named Kiley Hamlin, who's now an assistant professor at the University of British Columbia.

We created a set of one-act morality plays. For each of these, there is a character who tries to do something, and there's a good guy and there's a bad guy. These are animated figures or simple geometrical objects or puppets. For instance, in one of our studies, a character would be struggling to get up a hill. One guy would come and push him up. Another guy would come and push him down. In another, a character would be playing with a ball. He rolls the ball to another puppet. They look at each other and the puppet rolls it back. He rolls the ball to another puppet, they look at each other, and then this other puppet runs away with the ball. In a third one-act play, there's a puppet trying to open up a transparent box. The baby can see that there is a toy in there. And one puppet comes and helps to open the box and, later, a different puppet jumps on the box, slamming it closed.

These are three examples; we have a couple more scenarios in the works now. What we find is that if you ask toddlers of 19 months of age, "Who is the good guy?" and "Who is the bad guy?" they respond in the same way that adults do. They point to the proactive agent, the person who helps the character achieve his goals, as the good guy, and they point to the disrupter, the thwarter, as the bad guy.

Now, maybe that's not so exciting—these are fairly old kids. But what we've done is we've pushed the age lower and lower. In one set of studies, we present the baby with both characters, and we see where the baby will reach for, which one the baby will choose. Keep in mind that everything is counterbalanced, and the person who's offering the choice is always blind to the roles of the different characters, to avoid the problem of unconscious cuing. Also, the parents have their eyes closed during the study.

We find that down to six months of age, they'll reach for the good guy. We also have neutral conditions, and these tell us that they'd rather reach for the good guy than to a neutral guy, but they'd rather reach for a neutral guy than to a bad guy. This suggests that there are two forces at work—they are drawn toward the good guy and drawn away from the bad guy.

In a recent study that was just published in the journal *Developmental Science*, we test three-month-olds. Now, three-month-olds are blobs; they are meatloafs. They can't coordinate their actions well enough to reach. But we know from the six-month-old study that before babies reach, they look to where they're going to reach. So for the three-month-olds, we record where they look. And as predicted, they look to the good guy, not to the bad guy.

Does this show morality, a moral instinct? No. What it shows is that babies are sensitive to third-party interactions of a positive and negative nature, and this influences how they behave toward these characters, and, later on, how they talk about them. And I think that is relevant to morality. I think it's a useful moral foundation. But how moral is it? Are these truly moral judgments? And the honest answer is that we don't know. This is something we're actively exploring, but, as you could imagine, when you're dealing with six-month-olds, it's difficult to study.

We are embarking on some experiments that try to address this issue, along with a Yale graduate student, Neha Mahajan. One aspect of mature morality is that you not only approach a good guy and avoid a bad guy, but you believe that a good guy should be rewarded and a bad guy should be punished. So we tested 19-month-olds to see whether they share this intuition. Using our usual paradigms, we have a good guy and a bad guy and we ask the children to give a treat to one of them. And what we find is that they usually give it to the good guy. We also have a punishment condition, so we say to the child: you have to take a treat from one of these characters. They'll tend to take it from the bad guy.

Recently, with nine-month-olds, we did a study looking at their notions of justice. And to do this, we have a two-act play. In the first act, you have a good guy and a bad guy. And they do their good guy/bad guy actions, the ones that I described before. In the second act, what happens is that two more characters come in. In one condition, one of the characters rewards the good guy and the other character punishes the good guy. And we find that babies prefer, by reaching, the character who rewarded the good guy.

Now, this is not so surprising, because we had the previous finding that babies like positive actors. Maybe this is all that's going on. The second condition's more interesting. You have a good guy and a bad guy, then one character comes in and rewards the bad guy; another character comes in and punishes the bad guy. Now the babies robustly prefer the one who punishes the bad guy, suggesting that they will favor bad actions when they are done to those who are themselves bad. This suggests some rudimentary—and I'm happy to put into square quotes—some rudimentary sense of "justice."

There are other studies looking at baby morality from Renee Baillargeon's lab at University of Illinios and Luca Surian's lab from University of Trento, as well as from other labs. These also support the idea that there's both a surprisingly precocious grasp of moral notions and a surprisingly precocious propensity for moral action. Now, some would argue that I could stop my talk now because I've solved the problem I've set out to answer. The human niceness that we are interested in exists in babies; it is part of our hard-wired inheritance. We are, as Dacher Keltner put it, "born to be good." To the extent you find a narrowing of this kindness in adults, this is due to the corrupting forces of culture and society.

This is not the argument I wish to make. I find the idea of an innately pure kindness to be extremely implausible. For one thing, our brains have evolved through natural selection. And that means that the main force that shaped our psyche is differential reproductive success. Our minds have evolved through processes such as kin selection and reciprocal altruism. We should therefore be biased in favor of those who share our genes at the expense of those who don't, and we should be biased in favor of those who we are in continued interaction with at the expense of strangers.

Also, there is now a substantial amount of developmental evidence suggesting that this kindness that we see early on is parochial. It is narrow. It applies to those that a baby is in immediate contact with, and does not extend more generally until quite late in development.

Here are some sources of evidence for this claim. We've known for a long time that babies are biased toward the familiar when it comes to individuals. A baby will prefer to look at her mother's face than the face of a stranger. A baby will prefer to listen to her mother's voice as opposed to the voice of a stranger. This bias also extends to categories. Babies prefer to listen to their native language rather than to a language that's different from theirs. Babies who are raised in white households prefer to look at white people than at black people. Babies who are raised in black households prefer to look at black people than at white people.

We know that this last fact isn't because the babies know that they themselves are white or black, because babies that are raised in multiethnic environments show no bias. It has to do with the people around them. And as they get older, this bias in preference translates into a bias in behavior. Young children prefer to imitate and learn from people who look like them and those who speak the same language as them. Around the age of nine months, they'll show stranger anxiety—they avoid new people.

There are also studies now with preschool children, older children, and adolescents that show that it is fairly easy to get them to categorize in favor of their own group over others, even when the group is established in the most minimal and arbitrary circumstances. This is all based on Tajfel's work on "minimal groups." For instance, in experiments by Bigler and others, you take a bunch of children and you say okay, kids, I have some red T-shirts and blue T-shirts, I'm just going to give them to you guys. You get the children to put on the T-shirts, so that now you have a red T-shirt group and the blue T-shirt group. Now you approach a child from the red T-shirt group, and you say: I have some candy to give out, and you can't get any, but I'm asking you to give it to other people. Who do you want to give it to? Do you want to give it to everybody equally, or you want to give it more to the red or more to the blue?

It turns out that children are biased to give more to their own group, even when they don't personally profit from the giving. And when asked about the properties of their group—who's nice, who's mean, who's smart, who's stupid—a child who just put on a red T-shirt will tend to favor the red T-shirt group over the blue T-shirt group, even though it's perfectly clear that the assigned were divided on an arbitrary basis.

Yet another bit of bad news about human nature comes from economic games. Many of you are familiar with the ultimatum game, and this is just one of a series of games thought up by behavioral economists that purports to show niceness among adults, that we are generous in certain ways. Now, I am highly skeptical about what these studies really show, and we could talk about that in the question period. But what's interesting for these purposes is that children behave quite differently from adults in these games.

I'll give you one example. This is the dictator game. The dictator game is actually even simpler than the ultimatum game. I choose two people at random. One of them, the subject, is lucky. He gets some money, say $100. Now he can give as much as he wants to the other individual, everything from the entire $100 to nothing to all. This other person will never know who made the choice—it's entirely anonymous.

From a self-interested perspective, the subject should just keep all the money. But what you find is that people actually give. People give roughly 30 percent. Some people give nothing, but some people give half, and some people even give more than half. This is surprisingly nice. Ernst Fehr and Simon Gächter recently did this with children. What they did was set up a very simple ultimatum game. They gave children two candies. And they say to each child: you can either keep both candies or you can give one of them to this stranger.

Seven- and eight-year-olds will often choose to do a split. But younger children almost always keep both candies. So to the extent that there is generosity to strangers, it emerges late. Now, one problem with the standard ultimatum game is that you are pitting two impulses against each other. The child might have an equity/kindness/fairness impulse, so there is a desire to share, but the child might also like candies, so there's a desire to keep both

Paul Bloom

of them. You're pitting them against each other. Maybe children's hunger trumps their generosity.

Fehr and Gächter explored this with another study. The child got to choose between either getting a candy and giving another person a candy versus getting a candy and giving the other person nothing. Now, from a consequentialist point of view, this is not a head-scratcher. There are not two competing impulses. One can be nice without suffering any penalty. But until about the age of seven or eight, children are perfectly indifferent. The numbers are about 50 percent—they don't care. It's not like they hate the other anonymous person and want to deprive him of the candy, it is that they have no feelings either way.

This shouldn't surprise us. Maybe it's even better than we could have expected. The dominant trend of humanity has been to view strangers—nonrelatives, those from other tribes—with hatred, fear, and disgust. Jared Diamond talks about the groups in Papua New Guinea that he encountered. And he points out that for an individual to leave his or her tribe and just walk into another, strange tribe would be tantamount to suicide. Others have observed that the words that human groups use to describe themselves and others reflect this same animus toward strangers. So groups tend to have a word for themselves that often means something like person or human. Then they have a word for other people. And now sometimes this is just "The Others," like in the TV show *Lost*. But sometimes they describe the other group using the same word they use for prey, or food.

So there's a puzzle, then, because the niceness we see now in the world today, by at least some people in the world, seems to clash with our natural morality, which is nowhere near as nice. How did we end up bridging the gap? How have we gotten so much nicer?

Note that I'm focusing here on questions of our kindness to strangers, but this question could be asked about other aspects of morality, such as the origin of new moral ideas, such as that slavery is wrong or that we shouldn't be sexist or racist.

These are deep puzzles. I'll end this talk with two compatible theories of the emergence of mature human kindness.

The first involves increased interdependence. This is something that Robert Wright has been arguing in a series of books, and Peter Singer and Steven Pinker have also discussed it, in different forms. The idea is that as you come into contact with more and more people, in a situation where there is interdependence, where your life is improved by being able to contact with the other person, where there is a nonzero-sum relationship, you will come to care about their fates.

This is niceness grounded in enlightened selfishness. As Robert Wright once said in a talk, "One of the reasons I don't want to bomb the Japanese is that they built my minivan." Because he's in a commercial relation to these people, his compassion gets extended to where it wouldn't have otherwise been.

There is some support for this view, coming from a study by Joseph Henrich and his colleagues that was published in *Science* a few months ago. Henrich et al. looked at 15 societies, and they had the people in these societies play a series of economic games. They found considerable variation in how nice people are to anonymous strangers, and then did some analyses to see what determines this niceness. One finding is that capitalism makes people nicer. That is, immersion in a market economy has a significant relationship with how nice we are to anonymous strangers, presumably because if you're in a market economy, you're used to dealing with other people in long-term relationships, even if they're not your family and they're not your friends. The second factor was membership in a world religion—Christianity or Islam. This makes people nicer, perhaps because it immerses people into a larger social group and entrains them to deal with strangers.

Another explanation for the increase in human niceness is the power of stories. One of the consequences of fiction and of journalism is that they can bring distant people closer to you. You can come to think of them as if they are kin or neighbors. This can extend one's sympathies toward individuals, but, as Martha Nussbaum and many others have argued, it can also expand one's sympathy toward groups.

Consider moral progress in the United States. I think that the great moral change in our society over the last 50 to 100 years has been the changing attitudes of whites toward African-Americans. And the great moral change in the last ten years has been in straight people's attitudes toward gay people. I think that for both cases, the engine driving this change was not philosophical arguments or theological pronouncements or legal analyses, it was fiction. It was imagination. It was being exposed to members of these other groups in sympathetic contexts. I would argue, more specifically, that one of the great forces of moral change in our time is the American sitcom. I'll end by saying that this speaks to one of the issues that occupies many people in this room, which is the role of rational deliberation in morality. There seems to be a contradiction here. On the one hand, social psychologists have a million demonstrations that people are impervious to rational argument. And so the reason I've come to my views about slavery or gay people or whatever is most likely not because somebody gave me a real persuasive argument. On the other hand, we know full well that rational thought has made a difference in the world. Just as a recent example, Peter Singer's thoughts on issues such as how to treat nonhuman animals have changed the world.

I think one way out of this—and this is very similar to something that Jonathan Haidt has argued—is that reasons do affect us, but they do so indirectly, through the medium of emotions. If so, this suggests a research project of tremendous importance, one that asks: how do people come to have new moral ideas and how do they convey these ideas in ways that persuade others?

I've made three arguments here. The first is that humans are, in a very interesting way, nice. The second is that we have evolved a moral sense, and this moral sense is powerful, and can explain much of our niceness. It is far richer than many empiricists would have believed. But the third argument is that this moral sense is not enough, that accomplishments we see and we admire so much in our species are due to factors other than our evolutionary history. They are due to our culture, our intelligence, and our imagination.

David Pizarro

Psychologist, Cornell University.

Like the others, I'd really like to thank John and the Edge Foundation for bringing us out. I really feel like a kid in a candy store here, to be able to speak with everybody here on a topic that I actually thought was the nail in the coffin of my graduate career. But thanks to kind people, including Paul and Jon, it has not been the nail yet.

What I want to talk about is piggybacking off of the end of Paul's talk, where he started to speak a little bit about the debate that we've had in moral psychology and in philosophy, on the role of reason and emotion in moral judgment. I'm going to keep my claim simple, but I want to argue against a view that probably nobody here has (because we're all very sophisticated), but it's often spoken of emotion and reason as being at odds with each other—in a sense that to the extent that emotion is active, reason is not active, and to the extent that reason is active, emotion is not active. (By emotion here, I mean, broadly speaking, affective influences.)

I think that this view is mistaken (although it is certainly the case sometimes). The interaction between these two is much more interesting. So I'm going to talk a bit about some studies that we've done. Some of them have been published, and a couple of them haven't (because they're probably too inappropriate to publish anywhere, but not too inappropriate to speak to this audience about). They are on the role of emotive forces in shaping our moral judgment. I use the term "emotive" because they are about motivation and how motivation affects the reasoning process when it comes to moral judgment.

There are a variety of ways that emotional processes affect reason in a nuanced way, and I just want to briefly mention one way in which this is the case: we have some evidence for, in work I've done with some people here, and in work that John and his colleagues have done, showing that disgust sensitivity, just the simple *tendency* to experience an emotion, can actually on one account (an account that we believe although we don't have good causal evidence for) shape beliefs over time.

We've shown that disgust sensitivity, that is, people who are more likely to be disgusted, over time end up developing certain kinds of moral views. In particular, we've shown that not only are people more political conservative if they're more disgust sensitive, but they specifically are more politically conservative in the following ways: they tend to adhere to a certain kind of moral view that the conservative party in recent years in the United States has endorsed that's characterized by being against homosexuality and against abortion.

What we've shown is that people who are higher in disgust sensitivity, that is, people who are more easily disgusted, actually seem to have these views. One way of thinking of this is that early differences in emotional styles can actually shape the kinds of things that you find persuasive. It's not a simple case of an emotion influencing me, therefore my reason is shut off and I'm influenced by this gut reaction (which certainly happens). But it's a more complex view of the interaction.

What I'm going to talk about today is a different kind of interaction between emotion and reason, and I'm going to use the example of moral principles. By moral principles, I mean moral principles in the way that many people (normative ethicists, development psychologists) have used the term moral principles, which is to say a sort of guiding general principle that leads to specific judgments.

The reason we use moral principles, or at least the reason people think we ought to use moral principles, is that anytime we make a specific moral claim, like say murdering babies is wrong . . . maybe that one's too obvious . . . but any singular moral claim can't be defended—unlike empirical facts they can't simply be defended by, say, going to Wikipedia and saying, "See, here it shows that murdering babies is wrong."

It takes a different kind of evidence to back up a moral claim. There is no yardstick, there is no polling that you can do, although in some cases some people have argued there is evidence for certain kinds of moral claims. A lot of them require the defense that comes in the form of a principle. Moral principles provide justification, because you can't just say "Homosexuality is wrong." "Why?" "Well, *just because.*"

David Pizarro

Usually you provide some overarching principle. And the use of these overarching principles has been considered the hallmark of moral reasoning by, for instance, Lawrence Kohlberg and his colleagues. The reason that it seems so powerful is that it's a broader principle that's invariant across a whole bunch of similar situations. So it's an invariant principle, just like a broad mathematical principle would apply to specific problems.

I don't need to discuss this too much here, because it's already been explained and you probably already know, but two of the biggest, most popular moral principles that philosophers and psychologists have talked about are consequentialist defenses. So that is, if you have to use a rule, a general rule to determine what's right and what's wrong, you might say, well, what's right and what's wrong is solely determined by the outcome of any given act. So this one broad rule that one might call utilitarian (consequentialist is the more broad term).

Or you might say the way that we determine whether something is right or wrong is—and then here you would appeal to any of a number of deontological rules, such as that the murder of innocent people is always wrong, no matter what. Now, these two have been pitted against each other many times by philosophers. I should say they're not always in conflict, obviously. But the interesting cases have been when they *are* in conflict. Earlier Josh [Greene] talked to us about trolley problems as these sort of paradigmatic examples that you can use, the "fruit flies" of moral psychology, because you can easily look at people's responses to trolley problems and other similar dilemmas that pit consequentialism and deontological principles against each other.

The way that, traditionally, psychologists and philosophers have understood this is, you're confronted with a moral situation. You see a possible moral violation and you say, is there a principle that might speak as to whether this is morally wrong or not? And so you choose the appropriate moral principle, whether it's consequentialism or some sort of deontological rule, and you say aha! It applies here. And you say, therefore, this act is *wrong*.

What I want to present today is some evidence that this is, in fact, not the way we do it. Rather, what happens is there's a deep motivation we have to believe certain things about the world. And some of these things are moral things. We have a variety of moral views, specific moral claims that we all hold to be true, that we believe, first and foremost, independent of any principled justification. What happens is that we recruit evidence in the form of a principle, because the principle is so rhetorically powerful. Because it's convincing to somebody else, you can say, well, this is wrong because it causes harm to a large group of people.

I'm going to be borrowing from a large tradition in social psychology that says it's not that we don't reason; in fact, we reason a whole lot. It's just that the process of reasoning can go horribly awry because of a variety of biases that we have (for example, the confirmation bias that was mentioned earlier). There have been some classic studies in social psychology showing that we work really hard when we're presented with something that contradicts a view of ours.

One of my colleagues, who did some of these studies with me, Peter Ditto, has shown that if you tell people, look, there's this easy test that will tell you whether or not you have a disease. (He made up the disease, but let's just say it was herpes.) So let's say I present you with a little piece of paper, it's like a litmus strip. You put it in your mouth and then you dip it in a solution. If it turns green, he told some of the participants, that means that you're healthy. To other participants he said if it turns green, that means you're sick. Some participants he said if nothing happens, you're healthy. The other group was told that if nothing happens, you're sick. So for the participants who thought that if nothing happened, they were healthy—and by the way, the paper was inert; absolutely nothing ever happened—for the people who thought that nothing happening was a good thing, they dipped it in once and they're like, "Yes, I'm healthy!"

But for the people who were told that if nothing happens, you're sick, they were sitting there, doing this [illustrating dipping a paper repeatedly and checking each time]. This is a nice metaphor for what goes on in our mind when we're confronted with something that goes against one of our beliefs. We're very motivated to find whatever evidence we can that we're actually right.

One of the reasons I started thinking about this came from early experiences when I realized that when I would ask my parents certain things—I remember the first time I asked my parents about Hiroshima and Nagasaki. And I said, well, it kind of seems wrong. Right? So I said, you know, Mom, Dad . . . (I was 23) . . . no. Mom and Dad, this doesn't seem right. And so the answer they gave me was, well, if it weren't for us dropping the bomb on Hiroshima and Nagasaki the war would have lasted a lot longer, a lot more people would have died, and so it was the right thing to do—a very consequentialist justification.

Then when I would think about it, on other occasions they would reject this very same justification, by telling me "the ends never justify the means" when there was another completely different example from a different domain. But wait, the power of the principle is supposed to be that it applies in invariant fashion across all these examples, and here you've told me in one breath that consequentialism is the right moral theory, and in another breath you've told me that consequentialism is a horrible moral theory. So which one can it be?

So when I got the chance to actually do these studies, what

we did was we put together some studies that would test out this very idea—that what is going on is that people have a *moral toolbox*. That is, we are very capable of engaging in principled reasoning using deontological principles *or* consequentialist principles. What we do is simply pull out the right moral principle whenever it agrees with our previously held position.

This is contrary to what Josh [Greene] has shown on the face of it. But I don't think it's contrary at a deeper level. Often the tradition in moral philosophy and in moral psychology, when poring over these [moral dilemma] examples, is that they've always said, you know, "Jones and Smith," or "Person A and Person B," or "Bob and Alice," and what Josh has shown is that people have a default acceptance of deontological principles, but that if you get them to think hard enough, they'll often go with the consequentialist decision.

But what we were interested in, given the experience that I and my colleagues have had in actual moral discussions, was when it wasn't "Person A," "Person B," or "Jones and Smith"; it was actually your friend or your country or your basketball team, or something—situations that we engage in in everyday life, and which are infused with a motivation to favor your side, or to argue that you are right.

So we sought to design a set of experiments that would test whether or not this was actually the case. Would people appeal to or endorse explicitly a principle that justified what they believe to be true while claiming that it was invariant, while at the same time denying the same principle whenever the opportunity arose to criticize a moral view that they opposed?

What we did was try to find a natural source of motivation, and politics provides this wonderful natural source of motivation for moral beliefs, because so much of politics, as I think Jon [Haidt]

David Pizarro

has nicely pointed out, really is about moral differences, right? So we wanted to look at conservatives and liberals, their moral views and their appeal to these explicit moral principles, and whether or not they would really show the invariant sort of psychology that philosophers and psychologists would assume that we're supposed to use. Or whether they would actually just sort of use whatever principle justified their belief.

What we did was (to bring some trolleyology back into the mix), we used a modified version of the footbridge dilemma. As Josh described this morning, the footbridge dilemma is where five people are trapped on train tracks and there's one large person that you could push off to stop the train from killing the five. Most people find this morally abhorrent; they would never push the large person off the train. But we manipulated it a little bit. What we did was to give the large man a name. We told half of the people that they were in a situation where there were a large number of people that were going to get killed by a bus unless they pushed a very large man off of a footbridge.

We manipulated whether the guy's name was Tyrone Payton or whether his name was Chip Ellsworth III. We took this to be a valid manipulation of race, and (because reviewers ask for such things) it turns out that yes, most people think that Tyrone is black and most people think Chip is white. And so we simply asked people, would you push Chip Ellsworth III (or would you push Tyrone Payton)?

We also asked people to indicate [their political orientation] on a seven-point scale, where one meant they were liberal and seven meant they were conservative. In case you're curious, this simple one-item measure of liberal/conservative is correlated with much larger and more detailed measures of political orientation, and it predicts voting behavior, along with all the other things that you might want it to predict. We then asked people "was this action appropriate [i.e., pushing the large man off the bridge to save the greater number of people]," "do you agree with what the person did," and, critically, we asked people the general principle, we said, "do you agree with the following: sometimes it's necessary to kill innocent people for the sake of saving greater numbers of lives?" (It's a very, very straightforward principle.)

What we found was that liberals, when they were given Chip Ellsworth III, were more than happy to say, "clearly consequentialism is right." You push the guy, right? You've *got* to save the people. Self-reported conservatives actually reported the opposite—they were more likely to say that you should push Tyrone Payton, and that well, yes, consequentialism is the right moral theory (when the example was Tyrone Payton). When asked about the general principle [of consquentialism] they also endorsed the general principle.

We did this both at U.C. Irvine, and then we wanted to find a sample of more sort of, you know, real people, so we went in Orange County out to a mall and we got people who are *actually* Republicans and *actually* Democrats (not wishy-washy college students). The effect just got stronger. (This time it was using a "lifeboat" dilemma where one person has to be thrown off the edge of a lifeboat in order to save everybody, again using the names "Tyrone Payton" or "Chip Ellsworth III.") We replicated the finding, but this time it was even stronger.

If you're wondering whether this is just because conservatives are racist—well, it may well be that conservatives are more racist, but it appears in these studies that the effect is driven by liberals saying that they're more likely to agree with pushing the white man and disagree with pushing the black man. So we used to refer to this as the "kill whitey" study. It appears driven by a liberal

David Pizarro

aversion to killing, to sacrificing Tyrone, and not, in this case, to the conservatives. (Although if you want evidence that conservatives are more racist, I'm sure it's there.)

So we thought, okay, we demonstrated this using this traditional trolley example. Let's look at another moral example that might be more relevant and a bit more realistic. So this time we asked students (again at U.C. Irvine), we said here's a scenario in which there are a group of soldiers who are mounting an attack against the opposing military force (this is a classic "double effect" case from philosophy) where the military leaders knew that they would unintentionally kill civilians in the attack. They didn't want to, but they foresaw that it would happen.

For one set of respondents we described American soldiers in Iraq who are mounting an attack against Iraqi insurgents, and in this case, Iraqi civilians, innocent Iraqi civilians, would die. The other set read about Iraqi insurgents attacking American forces, and in this case, innocent American civilians would die.

What we found, again, was that when we asked people whether they were liberal or conservative, and we look at the split—I'll just give you the actual example we used: "Recently an attack on Iraqi insurgent leaders was conducted by American forces. The attack was strategically directed at a few key rebel leaders. It was strongly believed that eliminating these key leaders would cause a significant reduction in the casualties of American military forces and American civilians. It was known that in carrying out this attack, there was a chance of Iraqi civilian casualties. Although these results were not intended, and American forces sought to minimize the death of civilians, they did it anyway." And we tell them that sure enough, they do it and civilians die.

We then asked people, is this morally permissible? Is it okay to carry out a military attack when you unintentionally, but foreseeably, are going to kill civilians? And what you get, again, is a flip. This time it seems to be more motivated by a liberal bias for favoring the action of the Iraqi insurgents. So we used to refer to this as the "reasons for treason" study.

But you get a more natural crossover effect here, such that conservatives are more likely to say consequentialism is true, that sometimes innocent people just have to die, but only when it's the Iraqi civilians dying. Liberals are more likely to say consequentialism is true when American civilians are dying.

So across these studies, you know, we showed in a few trolley scenarios, and in this military action scenario, that in fact it's *not* that people have a natural bias toward deontology or a natural bias toward consequentialism. What appears to be happening here is that there's a motivated endorsement of one or the other whenever it's convenient.

Up until this point, though, we had relied on participants' pre-existing political beliefs, and we really wanted to see if we could manipulate it.

In the great tradition of social psychology, we primed participants with unscrambled sentences. Half of the participants got a task where they were supposed to unscramble sentences, and one of the words embedded in the sentence had to do with patriotism, and the other half got words that had to do with multiculturalism. We took this as a sort of proxy for manipulating political beliefs in the way that would be more closely aligned with "conservative" or "liberal."

What we found was, regardless of the political orientation participants came in with, these priming manipulations worked. In the exact same Iraqi/American military scenarios that I described earlier, if you were primed with patriotism, you showed the same pattern that conservatives showed, and if you were primed with

David Pizarro

multiculturalism, you showed the same pattern that liberals showed [in the previously described study].

So using a priming manipulation, in which we could embed the motivation in our participants—at least that's what we think we're doing—that motivation still causes differential endorsement of consequentialism across the different scenarios.

Here I just want to tell you that we've done this with a whole bunch of other scenarios, and that importantly, if you ask participants after they do the study whether or not race should play a role, whether nationality should play a role in these judgments, "do you think nationality or race should play a role in moral dilemmas?" overwhelming people say "no." In fact, I don't remember any one instance of somebody saying "yes." In fact, some people are so offended at us even intimating that they might be using race or nationality as a criterion that we get all kinds of insults, like "how dare you?" or "I'm not even answering this!" Another finding we got is that, if you're curious, that some people, when we presented these results, would say, well, you know, liberals are right because, in fact, a black life *is* worth more than a white life (these were white liberals in New York City, Chicago, just so you know). Fair enough, maybe that's the case for these people. But overwhelmingly, our participants, when asked, deny that this is the case.

Also, if, for instance, you give a liberal the scenario in which they're asked whether or not to sacrifice Tyrone, and they say "yes," and then you give them a scenario with Chip, they also say "yes." That is, they realize their inconsistency. We take this as evidence that people seem to have a desire for consistency—a sort of embarrassment at their own inconsistency.

This is not just about consequentialism and deontology; these just provide a nice set of examples of moral principles that other psychologists use. But we've also asked, just to give you an example, we asked people about freedom of speech: do you think that freedom of speech is a principle that should be upheld no matter what? Depending on the scenarios that you give, for instance—so to half of the participants we said there's a Muslim protester that was burning the American flag and proclaiming America to be evil.

When we asked the liberals and conservatives, "Do you agree with what he did?" most people say "no." When we ask, "Do you think that freedom of speech should be protected despite this?" liberals say, "Of course freedom of speech should be protected!"; conservatives say, "Absolutely not, this is just not right." But if you give participants a scenario in which a Christian protestor showed cartoons of Mohammed and proclaimed that Islam was evil, the conservatives say, "Freedom of speech should be protected, no matter what!" and the liberals say, "No way, freedom of speech has its limits!"

Again, some other scenarios that we used included asking people if it was okay to break a law if you perceived it be unjust; in one case it was pharmacists who gave the "morning after" pill, despite it being illegal, and in another case it was [doctors refusing to aid in capital punishment]. Here again, you can get a liberal/conservative split in the endorsement of the principle of [breaking a law if it is believed to be unjust].

Across a whole body of studies, I think what we've shown is that it's not that reasoning doesn't occur, and in fact, reasoning can actually be persuasive. Say we're in a social situation and I want to convince you of something, and I say it's wrong to do X. And you say, Well, why is it wrong? I appeal to a broader principle. And this, in fact, might work. This might convince you, oh yes, that's a good point, that *is* a principle that we should adhere to. It's just that we're sneaky about it, right?

David Pizarro

And it's not that motivation [to uphold our moral beliefs] is wiping out our ability to reason, it's just making it very directional. If anything, it's opening up this skill and ability we have to find confirmation for any belief that we might have. This is probably true in most domains of social judgment, but I think it's especially interesting in the domain of moral judgment because of the implications that this has.

I want to say, before I end, that I've been a champion of reason, and I've often said that rationality can actually change moral beliefs. In fact, early on, after Jon [Haidt] published his widely read and influential paper on emotions being in charge of the moral domain, I said "Absolutely not!" Paul [Bloom] and I had a little paper where we said "Absolutely not, you're wrong!" But ever since then I appear to have done nothing but studies supporting Jon's view.

Even then, I am still an optimist about rationality, and I cling to the one finding that I talked about, which is that when you point out people's inconsistencies, they really are embarrassed. Hopefully, at a very minimum, what we can say is, let's at least keep pointing out the stupidity of both liberals and conservatives. And at this point, I'll come out and say that, Jon, I also am not someone who is a self-proclaimed liberal, although I'm sure on any test I would appear to be liberal.

But I, in fact, am a libertarian communist, and oftentimes am quite apathetic about politics. And I think this is what allows me to do these studies with the impartiality of a scientist.

Joshua Knobe

Associate Professor, Program in Cognitive Science and the
Department of Philosophy, Yale University.

So far we have been talking about questions in moral psychology. So we've been talking about the questions: How is it that people make moral judgments? Do they make moral judgments based on emotion or reason? Is it a capacity that's just learned or is it something that's innate?

In this last talk, I thought I'd take things in a slightly broader direction. What I want to ask about is a slightly different question: What is the role of people's moral thinking in their cognition as a whole? What is the role of this particular type of thinking—morality—in our entire cognition, the way we think about things in general?

Suppose we look out at a certain kind of situation unfolding. As we look at this situation, we might make certain types of moral judgments about it. We might think that a person in this situation is doing something morally wrong, or morally right. We might think that the person deserves praise or blame for what the person is doing. But we might also think about all sorts of other aspects of the situation. We might look at a situation and just wonder: What are these people doing? Or we might wonder: What do they intend to accomplish? Or we might wonder: Are these people happy or unhappy? What are they going to be causing down the line and so forth?

So the question now arises: What is the relationship between these two aspects of people's thought? What is the relationship between the way they think about moral questions (like, say, whether someone did something morally wrong or morally right) and the way they think about these straightforward, factual questions? Questions like: What are these people doing? What are they intending to accomplish? What are they are going to be causing? And so forth. What is the relation between these two kinds of judgments?

And here there's been a kind of a traditional picture that long dominated the field. The picture was that one could think about the relationship between these two things in terms of something like a series of stages. So you look out on the situation. The first thing you do is you just try to figure out what is actually going on in this situation. So you're just trying to think: What are these people doing? What are they intending? What are they causing? Just try to get a grip, in a purely factual, broadly scientific way, on what's occurring in this situation. And then once you figure that out—once you really know what's happening in the situation—then you can do something further, as a kind of second stage. In the second stage, you take in all this information about what these people are doing, and you use it to make a further judgment, a moral judgment about, say, whether what they're doing is morally right or morally wrong.

So in this kind of view, moral cognition occupies just a small and very delimited aspect of our cognition as a whole. When we try to understand the world, most of what we're doing has nothing to do with morality. It's just this purely scientific attempt to understand things. And then there's this kind of extra little step at the end. After everything else is done, this extra little step where we engage in moral cognition and try to make a moral judgment. So, on this view, moral cognition doesn't end up being that important, but it is playing a certain role in one aspect of the process.

I think this kind of view is somehow deeply intuitive on a certain level. There's something that really resonates with people about this kind of picture. For that reason, it has long held a grip on the whole way that the study of moral cognition worked.

But just in the past few years there's been a growing challenge to this view. This challenge has come from a kind of surprising source. It happened that a group of people who are actually in philosophy departments began thinking that the right way of going about understanding concepts would be to actually do experimental studies about how people use these concepts. These were philosophers just trying to understand the concept of knowledge, the concept of causation, the concept of intention, and so forth. And they began thinking: If we really want to understand how these concepts work, we can't just sort of sit here in our armchairs trying to figure it out by reflecting on it. Maybe we should go out and actually do experiments to see how could we use these concepts. So we're going to do studies in which we systematically vary particular factors and then show how varying those factors will influence people's application of these ordinary concepts.

This really ended up being a kind of revolutionary move within philosophy. It led to a sort of new and different approach to engaging in philosophical inquiry. Just as a number of years ago, there was a time in which economists first began doing economic experiments and this gave birth to this field of experimental economics, so too now there's been this real shift toward this new idea of what's called *experimental philosophy*.

But what's really exciting about this new work is not so much just the very idea of philosophers doing experiments but rather the particular things that these people ended up showing. When these people went out and started doing these experimental studies, they didn't end up finding results that conformed to the traditional picture. They didn't find that there was a kind of initial

Joshua Knobe

stage in which people just figured out, on a factual level, what was going on in a situation, followed by a subsequent stage in which they used that information in order to make a moral judgment. Rather they really seemed to be finding exactly the opposite.

What they seemed to be finding is that people's moral judgments were influencing the process from the very beginning, so that people's whole way of making sense of their world seemed to be suffused through and through with moral considerations. In this sense, our ordinary way of making sense of the world really seems to look very, very deeply different from a kind of scientific perspective on the world. It seems to be value-laden in this really fundamental sense.

So I thought what I'd do today is just to talk about a couple of examples in which this sort of effect seems to be arising and then get your input about how we might be able to understand it. The very first experiment that I'm going to be describing is one that I think a lot of you already know. But then after that, we're going to be talking about newer work that I think most of you won't already be familiar with.

The very first effect I want to mention is an effect on people's ordinary intuitions about the concept of intentional action. This is the concept we use to distinguish between things that people do intentionally, like, say, taking a sip of wine, and things that people do unintentionally, like, say, spilling the wine all over one's own shirt. So the question now is: How does this distinction work? How do people ordinarily distinguish between things that are done intentionally and those that are done unintentionally?

When you first begin to consider this, there's a sort of view that seems really tempting. And the view is: this distinction is just a straightforward, factual distinction. It just has to do with the mental states of the person performing the action. What it is to do something intentionally is something like: to know that you're doing this thing, or perhaps to want to do this thing, something like that.

But we started thinking, maybe there's actually more to this case. Maybe there was something more subtle going on in people's intuitions about intentional action. And, in fact, maybe people's *moral* judgments were actually playing a role in their notion of what it is to do something intentionally or unintentionally.

The thought then was, if you want to know whether someone did something intentionally or unintentionally, it's not enough just to know what they wanted, or what they knew—you have to make a judgment about whether what they are doing is something morally bad or something morally good.

To get at this, initially, we just ran a very simple study in which participants were assigned to one of two conditions. So there were two different conditions, and participants were randomly assigned to them. They differed in just one respect.

Participants in one condition got the following story. The story went like this.

The vice president of a company goes to the chairman of the board, and he says, "Okay, we've got this new policy. It's going to make huge amounts of money for our company. But it's also going to harm the environment." And the chairman of the board says, "Look! I know this policy's going to harm the environment. But I don't care at all about that. All I care about is making as much money as we possibly can. So let's go ahead and implement the policy." They implement the policy and then sure enough, it ends up harming the environment.

And the question then was: Did the chairman of the board harm the environment intentionally?

Faced with this question, most people say, "Yes!" They say the

Joshua Knobe

chairman of the board harmed the environment intentionally. And when you first start thinking about why they might say yes to a question like this, it seems like it probably just has something to do with the mental states that this chairman is described as having. So the chairman knows that he's going to harm the environment, and he goes ahead and does it anyway. Maybe it's that fact, the fact that he had this particular mental state, that makes people think he did it intentionally.

But we were thinking maybe there's actually something more complex about it. Maybe part of the reason that people say he did it intentionally is not just his knowledge, or the mental state he had, but the fact that they make this particular moral judgment, the fact that they judge that harming the environment is something morally bad and morally wrong for him to do. So participants in the other condition got a case that was almost exactly the same except for one difference, which is that the word "harm" was changed to "help." So the story then becomes this:

The vice president of a company goes to the chairman of the board and says, "Okay, we've got this new policy. It's going to make huge amounts of money for our company and . . . it's also going to help the environment." And the chairman of the board says, "Look! I know this policy is going to help the environment but I don't care at all about that. All I care about is just making as much money as we possibly can. So let's go ahead and implement the policy." So they implement the policy. And sure enough, it helps the environment.

And now the question is: Did the chairman help the environment intentionally?

But here people don't give the same response. They don't say the chairman helped the environment intentionally. Instead they seem to say that the chairman helped the environment unintentionally.

But look at what's happening in these cases. In the two cases, it seems like the chairman's attitude is exactly the same. In both cases, he knows the outcome is going to occur, he decides to do it anyway, but he doesn't care about it at all; he's not trying to make it happen. The thing that's differing between the two cases is just the *moral* status of what the person is doing. In one case he's doing something bad, harming the environment, and in the other case, he's doing something good, helping the environment. So it seems somehow that people's moral judgments can affect their intuitions just about whether he did it intentionally or unintentionally.

When we first came out with these results a number of years ago, it was really puzzling why this might be happening. We really couldn't understand what might be going on here, what might be the boundaries or the nature of the effect. But just in the past couple of years, a whole bunch of different studies by different researchers are really giving us a lot of insight into why this effect is occurring, what might really be going on here. I thought that a good way of discussing some of this work would be to talk about the work on this that has been done by the people here, the people at this workshop. And maybe that could also just give a general sense of how work on this topic has been proceeding.

I thought I'd start by talking about a study that was done in Marc Hauser's lab. Marc Hauser was interested in the question: Is this effect, the effect that we just described, due to people's emotional responses? Is it something about the way we're emotionally responding in these cases that leads us to say that the harming is intentional but the helping is unintentional?

The researchers in his lab came up with a really interesting way of going about testing this question, and that was to look at people who have really serious deficits in the capacity for emotional response. If we want to know why ordinary people respond the way

Joshua Knobe

they do, whether the ordinary people's responses is due to the sort of emotional reaction they have, we can go to people who differ from us in the relevant respect, who have a real impairment in the capacity for emotional response and then see how they respond. Hauser's lab looked at the intuitions of people who have lesions to the ventromedial prefrontal cortex.

As a number of people at this workshop so far have mentioned, people who have these lesions, VMPFC lesions, show really, really severe impairments in the capacity for emotional response. Their capacity to respond emotionally is really different from the way that normal subjects would. And on moral questions that have been claimed to involve emotion, they give really, really different responses. So a team of researchers, led by this pretty amazing graduate student named Liane Young, went to these participants, participants with lesions to the ventromedial prefrontal cortex, and gave them that exact case that I just described—the case of the guy who either harms the environment or helps the environment.

The results came out as something of a surprise. These participants showed exactly the same asymmetry we find among normal participants. Even these people with really serious impairments in the capacity for emotional response still said that he harmed the environment intentionally but that he helped the environment unintentionally. So what Liane Young, Marc Hauser, and their colleagues then concluded is that maybe this effect actually has nothing to do with emotion. Maybe it's not due to an emotional response we're having. Maybe we really have a kind of purely cognitive system for a kind of a nonconscious moral appraisal, and it's that purely cognitive, nonconscious appraisal that's driving the effect we find in normals.

So we have this emotional response to these cases. But if you take that emotional response away from us, we still show the same answer. So it must not have been the emotional response that was driving our answers in the first place.

But, of course, in cases like these, we often find different experimental studies that push in different kinds of directions. And this case is no exception. It happens that David Pizarro and Paul Bloom also ran a series of studies on this topic, but their study ended up moving in a really different direction. They were thinking maybe people's emotional responses *do* have an impact on these judgments. In particular, they were thinking maybe people's emotional responses could have an impact even in cases where they don't endorse those emotional responses.

Suppose you look at a situation. You see someone doing something, and you find what this person is doing really disgusting. It disquiets you or disturbs you. You feel sort of bad about it. But then suppose you reflect. You reason about the situation. You conclude that what this person is doing is not actually bad—it's not wrong at all. So what you think, ultimately, is that there was no reason to be disgusted by it. Your disgust is misplaced. It's actually a perfectly fine action.

Their suggestion was that in a case like that, you would find the effect emerging anyway. The immediate emotional response people have would still impact their judgments about whether the person acted intentionally, even in cases where they didn't endorse that sort of emotion.

What they needed then was a case in which people would have an immediate emotional response that they wouldn't end up endorsing. So how do you find such a case? Well, one of the cases they came up with was the case of interracial sex.

They began by just asking participants straight up. "What do you think of interracial sex? Is there anything wrong with interracial sex?" You'll be happy to know that 100 percent of subjects say,

"Absolutely not. There is nothing, nothing wrong with interracial sex. It's completely, completely fine."

But then, they had a kind of trickier way of getting at people's intuitions about this question. And the way was this. They presented a case that was much like the one that I described earlier with the chairman. Subjects were told:

Imagine there's an executive at a record company. He's meeting with his assistant, and his assistant says, "Okay, we've got this new music video. It's really going to increase sales of this album, but we've been looking at the images in this video, and we think it's also going to encourage interracial sex. In particular, it's going to encourage sex between black men and white women."

The record executive thinks about it for a moment and then says, "Look! I know that the images in this video are going to encourage interracial sex, but I don't care at all about that. All I care about is increasing sales of this album. So let's go ahead with this new video. We're going to release it." They release the new video, and sure enough, it ends up encouraging interracial sex.

And now the question is: Did he encourage interracial sex intentionally?

What you see here's a really striking kind of correlation. Those participants who were high in a dispositional tendency to feel the emotion of disgust—those participants who just, in general, about various events that might occur to them in their lives, feel higher levels of this emotion of disgust—tend to say that he intentionally encouraged interracial sex. Whereas those participants who are low in the dispositional tendency to feel disgust, they say that he unintentionally encouraged interracial sex.

So what we see here is this striking finding. Even though all these participants swear that they find nothing wrong with interracial sex, those who had higher levels of disgust have a kind of immediate emotional reaction to this case (which they choose not to endorse), and maybe that immediate emotional reaction is still impacting their intuitions about whether it's done intentionally or unintentionally.

Looking at these different kinds of data, it's very difficult to figure out what to make of them. Maybe further studies can be done that may help disentangle these kinds of effects, but instead of disentangling them right now, let's turn to a completely separate example, a completely separate kind of case in which a similar sort of effect arises.

This is the concept of happiness. The concept of happiness is the concept we use when we ask, "Is this person truly happy?" Or when we ask, "Has this person really achieved happiness?" And the question now arises: What is that concept? What is our concept of happiness?

Initially, you might think exactly the same thing that people thought about intentional action. You might think that the concept of happiness is just the concept of having a certain kind of mental state. So to have happiness is to have these emotions: feelings of pleasure, enjoyment in the things that you're doing, a sense of satisfaction. If you have those mental states, then you're happy. If not, then you're not happy. It's just a purely scientific kind of thing.

But then, as we started thinking about it, we starting thinking: Maybe there's actually something more to it. Maybe when you pick out a person and you ask "Is this person truly happy?" you're not just asking, "Does this person have this psychological state?" You're asking something more, something that involves a certain kind of value judgment. You're asking: Does this is person have a life that has a certain kind of value or meaning in some way?

So if you say "I hope that this child will grow up to be happy," you're not just saying, "I hope this child will grow up to have a

Joshua Knobe

certain psychological state." You're saying, "I hope this child will grow up to have a certain kind of life, a life about which you can make a certain kind of value judgment."

If that's right, then the concept of happiness is really different from other kinds of concepts we use to understand the emotions. In particular, it would even be different from the concept of *unhappiness*.

Before I talk about any of the data, let me just see if you can kind of get this picture intuitively. Suppose we're talking about someone, and I say, "Is she truly happy?" The intuition you're supposed to have is that when I ask that question, I'm not just asking, "Does she have a certain psychological state?" or "Does she feel certain emotions?" "Is she feeling a kind of pleasure?" We seem to be getting at something more. When you say she's truly happy, you're saying that her life has some kind of meaning or value.

But now consider the opposite question. Suppose we're talking about someone, and I say, "She's truly unhappy." In saying "She's truly unhappy," I'm not saying that her life actually is bad or empty of meaningless. It seems like I'm just saying that she has a particular kind of emotion. She has a particular kind of feeling. A feeling of unhappiness. So we were thinking that these concepts might differ in a sort of systematic way.

Before I talk about this difference between happiness and unhappiness, I'll just give you a feeling for the kind of effect you see just within the case of happiness. (This is a study that's definitely unpublished—we just submitted it a couple of weeks ago.) In these studies, we just asked people questions. We presented people with vignettes that differ in certain respects and just asked whether a person is happy.

So subjects in the first case got the following condition.

Maria is the mother of three children who all really love her. In fact, they couldn't imagine having a better mom. Maria usually stays pretty busy taking care of her children. She often finds herself rushing from one birthday party to the next and is always going to pick up some groceries or buy school supplies.

While Maria has been preoccupied with her children, she does get to see her old friends occasionally. Almost every night she ends up working on some projects for the next day, or planning something for her children's future.

So here, participants are given this story about a really meaningful life that has this really wholesome objective. Then participants are told that the person who has this life has these really positive emotional states. So these participants are told:

Day to day, Maria usually feels excited and really enjoys whatever she's doing. When she reflects on her life, she also feels great. She can't think of anything else in the world that she would spend her time doing and feels like the success she's had is definitely worth whatever sacrifices she's made.

And these participants were just asked a simple question: Is Maria happy?

Then participants in the other condition, just as in the story of the chairman who helps or hurt the environment, got a story of someone who has a very different kind of life, a life that was different from a value-laden perspective, a person whose life is sort of vapid, or meaningless or empty in a certain respect. So here's the story that the participants in the other condition got:

Maria wants to live the life of a celebrity in L.A. In fact, she's even started trying to date a few famous people. Maria usually stays pretty busy in trying to become popular. She often finds herself rushing from one party to the next and is always going to pick up some alcohol or a dress. Maria is so preoccupied with becoming popular that she's no longer concerned with being honest to her old friends unless they know someone famous. Almost every night she ends up either drunk or doing some kind of drug, just like the famous people she wants to be like.

But then participants in the second condition were then given the information that she has really positive emotional states. In fact, they were given exactly the same paragraph as participants in the first condition:

Maria usually feels excited and really enjoys whatever she's doing. When she reflects on her life, she also feels great. She just can't think of anything else in the world that she would spend her time doing and feels like the success she's had is definitely worth whatever sacrifices she had made.

And these participants were then given the same question. Is Maria happy?

So participants in the two conditions are given exactly the same information about what's going on in Maria's head. She has all these kind of positive emotions, feels all this pleasure, takes a lot of enjoyment in her life. But they were given really different information about whether her life is kind of valuable or meaningful one.

It turns out that they make, correspondingly, really different judgments as to whether she's happy. Even though she experiences positive emotions in both cases, people say that she is happy when her life is a meaningful and valuable one but she's not happy in the case where it's not.

So we ran a number of different studies, trying to test for this, controlling for various variables. We always found the same effect emerging. If people believe that Maria is having a sort of a vapid or empty or meaningless life, they're unwilling to say that she is happy.

But, of course, the real claim I wanted to make was not just about this one concept, the concept of happiness, but about the idea that this concept differs from other concepts because the concept of happiness is different from other kinds of emotion concepts in a particular way. So to get at this notion, we ran a separate series of studies that used a 2x2 table. So we have a 2x2 design in which we independently manipulate two different factors.

First of all, we have a case in which someone has a valuable life (like, say, being a mother) or a case in which someone has this nonvaluable, or a bad life (like, say, being this Paris Hilton kind of character).

Then, separately from that, we have this other difference between different conditions. Some participants were asked the question: "Is Maria happy?" and other participants were asked the question: "Is Maria unhappy?"

If you look at the judgments for whether Maria is happy, they just replicate the existing finding. People say that Maria is happy if she has a good life, but they do not say that Maria is happy if she has a bad life. But now suppose you just turn to this other question. You just ask subjects: "Is she unhappy?" There, you see *absolutely no effect*. There's no effect of people's moral judgments on whether people say that Maria is unhappy. The only thing that affects whether they say that she's unhappy or not is just what mental state she's described as having. Is she described as feeling bad? Or is she described as feeling good?

Across three different studies, we found the same interaction pattern arising. For judgments about the question "Is she happy?" the moral status of her life is playing a big role. For judgments of the question "Is she unhappy?" there is no effect of moral judgment.

In conclusion, I just want to say a few brief words about where all this seems to be heading. When we first started out doing this kind of research, it seemed like maybe the effect we were seeing for the concept of intentional action was just some weird kind of quirk or idiosyncrasy, something kind of bizarre about this one concept. So we thought that there was something special going on about this one concept, the concept of intentional action, but that the kind of traditional picture I started out by describing might still be right.

But as research on these kind of phenomena has proceeded, we seem to be finding this same kind of effect again and again for every single concept that we're looking at. So it's not just that moral judgments impact people's intuitions about intentional action. It's not just that they impact intuitions of our happiness. They also impact intuitions about knowledge, about causation, about freedom, even about whether a particular trait counts as being innate.

What it's really starting to look like now is that the initial effect we saw for the concept of intentional action was really just a symptom of a much broader phenomenon whereby people's moral judgments seem to be infecting their whole way of understanding the world they see around them. I'd be really interested in hearing what folks here think about how this phenomenon might be explained.

16

The Marvels and the Flaws of Intuitive Thinking: Edge Master Class 2011

Daniel Kahneman

Recipient of the 2002 Nobel Prize in Economics; Eugene Higgins Professor of Psychology Emeritus, Princeton University; author, *Thinking, Fast and Slow.*

The marvels and the flaws that I'll be talking about are the marvels and the flaws of intuitive thinking. It's a topic I've been thinking about for a long time, a little over 40 years. I wanted to show you a picture of my collaborator in this early work. What I'll be trying to do today is to sort of bring this up-to-date. I'll tell you a bit about the beginnings, and I'll tell you a bit about how I think about it today.

Amos Tversky (1937–1996)

This is Amos Tversky, with whom I did the early work on judgment and decision making. I show this picture in part because I like it, in part because I very much like the next one. That's what Amos Tversky looked like when the work was being done. I have always thought that this pairing of the very distinguished person and the person who is doing the work tells you something about when good science is being done, and about who is doing good science. It's people like that who are having a lot of fun, who are doing good science.

We focused on flaws of intuition and of intuitive thinking, and I can tell you how it began. It began with a conversation about whether people are good intuitive statisticians or not. There was a claim at the University of Michigan by some people with whom Amos had studied, that people are good intuitive statisticians. I was teaching statistics at the time, and I was convinced that this was completely false. Not only because my students were not good intuitive statisticians, but because I knew I wasn't. My intuitions about things were quite poor, in fact, and this has remained one of the mysteries, and it's one of the things that I'd like to talk about today—what are the difficulties of statistical thinking, and why is it so difficult?

We ended up studying something that we call "heuristics and biases." Those were shortcuts, and each shortcut was identified by the biases with which it came. The biases had two functions in that story. They were interesting in themselves, but they were also the primary evidence for the existence of the heuristics. If you want to characterize how something is done, then one of the most powerful ways of characterizing the way the mind does anything is by looking at the errors that the mind produces while it's doing it, because the errors tell you what it is doing. Correct performance tells you much less about the procedure than the errors do.

We focused on errors. We became completely identified with the idea that people are generally wrong. We became like prophets of irrationality. We demonstrated that people are not rational. We never liked this, and one of the reasons we didn't was because we were our own best subjects. We never thought we were stupid, but we never did anything that didn't work on us. It's not that we're studying the errors of other people; we were constantly studying the way our own minds worked, and even when we knew better, we were able to tell what were the mistakes that were tempting to us, and basically we tried to characterize what were the tempting mistakes.

That was 40 years ago, and a fair amount has happened in the last 40 years. Not so much, I would say, about the work that we did in terms of the findings that we had. Those pretty much have stood, but it's the interpretation of them that has changed quite a bit. It is now easier than it was to speak about the mechanisms that we had very little idea about, and to speak about, to put in balance the flaws that we were talking about with the marvels of intuition. We spent most of our time on flaws because this was what we were doing. We did not intend the message that was in fact conveyed that the flaws are what this is about. In fact, this is something

Daniel Kahneman

that happens quite a lot, at least in psychology, and I suppose it may happen in other sciences as well. You get an impression of the relative importance of two topics by how much time is spent on them when you're teaching them. But you're teaching what's happening now, you're teaching what's recent, what's current, what's considered interesting, and so there is a lot more to say about flaws than about marvels. Most of the time everything we do is just fine, but how long can you spend saying that most of what we do is just fine? So we focused a lot on what is not fine, but then the students and the field get the impression that you're only interested in what is not fine. That is something that happened to us fairly early. We didn't do enough to combat it. But there was very little we could do to combat it, because most of our work was about flaws. In recent years it is becoming more even. We understand the flaws and the marvels a little better than we did.

One way thoughts come to mind

24 x 17 = 408

Another way thoughts come to mind

is angry

One of the things that was was not entirely clear to us when we started, and that has become a lot clearer now, was that there are two ways that thoughts come to mind. One way a thought can come to mind involves orderly computation, and doing things in stages, and remembering rules, and applying rules. Then there is another way that thoughts come to mind. You see this lady, and she's angry, and you know that she's angry as quickly as you know that her hair is dark. There is no sharp line between intuition and perception. You perceive her as angry. Perception is predictive. You know what she's going to say, or at least you know something about what it's going to sound like, and so perception and intuition are very closely linked. In my mind, there never was a very clean separation between perception and intuition. Because of the social context we're in here, you can't ignore evolution in anything that you do or say. But for us, certainly for me, the main thing in the evolutionary story about intuition is whether intuition grew out of perception, whether it grew out of the predictive aspects of perception.

If you want to understand intuition, it is very useful to understand perception, because so many of the rules that apply to per-

Daniel Kahneman

ception apply as well to intuitive thinking. Intuitive thinking is quite different from perception. Intuitive thinking has language. Intuitive thinking has a lot of world knowledge organized in different ways than mere perception. But some very basic characteristics that we'll talk about of perception are extended almost directly into intuitive thinking.

What we understand today much better than what we did then is that there are, crudely speaking, two families of mental operations, and I'll call them "type one" and "type two" for the time being because this is the cleaner language. Then I'll adopt a language that is less clean, and much more useful.

Type one is automatic, effortless, often unconscious, and associatively coherent, and I'll talk about that. And type two is controlled, effortful, usually conscious, tends to be logically coherent, rule-governed. Perception and intuition are type one—it's a rough and crude characterization. Practiced skill is type one, that's essential, the thing that we know how to do like driving, or speaking, or understanding language and so on, they're type one. That makes them automatic and largely effortless, and essentially impossible to control.

Type two is more controlled, slower, is more deliberate, and as Mike Gazzaniga was saying yesterday, type two is who we think we are. I would say that if one made a film on this, type two would be a secondary character who thinks that he is the hero because that's who we think we are, but in fact, it's type one that does most of the work, and it's most of the work that is completely hidden from us.

I wouldn't say it's generally accepted. Not everybody accepts it, but many people in cognitive and in social psychology accept that general classification of families of mental operations as quite basic and fundamental. In the last 15 years there has been huge progress in our understanding of what is going on, especially in our understanding of type one thought processes. I'd like to start with this example, because it's nice.

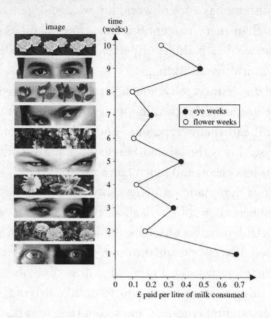

This is a report of the study that was carried out in a kitchen at some university in the UK, and that kitchen has an honesty system where people put in money when they buy tea and milk, and this is per liter of milk consumed, and somebody had the bright idea of putting a poster right on top of where the milk and the tea are, and of changing the poster every week. You can see that the posters alternate week by week, they alternate flowers and eyes, and then you can see how much people are paying. It starts from the bottom, which is the biggest effect, and this thing speaks for itself. It shows the enormous amount of control that there is, a thing that people are completely unaware of. Nobody knew that the posters had anything to do with anything. The posters are eyes, the eyes are symbols, somebody is watching, and that has an effect and you contribute more.

Daniel Kahneman

The power of settings, and the power of priming, and the power of unconscious thinking, all of those, are a major change in psychology. I can't think of a bigger change in my lifetime. You were asking what's exciting? That's exciting, to me.

This is an extreme case of type one. It's completely unconscious and very powerful, mixes cognition and motivation in a way that is inextricable and has a lot of control over behavior.

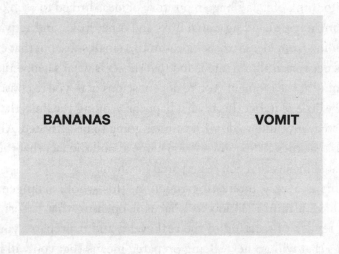

Let me propose what's happened in your minds over the last few seconds as you were looking at this display of two words. Everything that I'm going to say is very well-defended in recent research, mostly in research over the last 10, 15 years. You saw those two words. People recoil from the word "vomit." You actually move backward, and that has been measured. You make a face of disgust. It's not very obvious, but there is a disgust face being made. You feel a bit bad. The disgust face makes you feel worse because we know that forcing, shaping people's face into a particular expression changes the way they think and the way they feel. All of this happened. You are now prepared. Then something else happens.

You have those two words that have nothing to do with each other; you made a story. There is now a connection between those two, the banana has somehow caused the vomit, and temporarily you're off bananas. They're sort of past. This is not something you want.

Quite interesting. What happens is that your associative structure, your associative memory (and I'm going to be speaking mostly of what happens in associative memory), has changed shape. You have to think of it as a huge repository of ideas linked to each other in many ways, including causal links and other links, and activation spreading from ideas to other ideas until a small subset of that enormous network is illuminated, and that subset is what's happening in the mind at the moment. You're not conscious of it, you're conscious of very little of it, but this is what happens. Among the links that are activated, anything to do with vomit is going to be activated. All the words, "sickness," "hangover," you name it, and the fact that they're activated means that you're prepared for them.

This is a very interesting function, this whole arrangement. You have a mini-reaction to what is happening that is sort of a faint replica of reacting to the real event, and it prepares you for things that will go next. Being prepared means that you will need less stimulation in order to recognize a word, so that's the main thing. The amount of energy that it takes for a word to register is going to be sharply reduced for all the words that have to do with "vomit," and many of the words that have to do with "banana."

Links are going to be created, and in particular, and that's a fascinating part of it. You made up a causal story—that is, the vomit, which looks like an effect here, is looking for causes, and so you are making up scenarios. All of this is happening involuntarily, most of it is happening unconsciously, but you are primed, and you are ready for what comes next after that—particular stories of bananas causing vomit.

Daniel Kahneman

One other thing that I find most striking about this whole story that we've learned in the past 10, 15 years is what I call "associative coherence." Everything reinforces everything else, and that is something that we know. You make people recoil; they turn negative. You make people shake their heads (you put earphones on people's heads, and you tell them we're testing those earphones for integrity, so we would like you to move your head while listening to a message, and you have them move their head this way, or move their head that way, and you give them a political message)—they believe it if they're doing "this," and they don't believe it if they're doing "that." Those are not huge effects, but they are effects. They are easily shown with a conventional number of subjects. It's highly reliable.

The thing about the system is that it settles into a stable representation of reality, and that is just a marvelous accomplishment. That's a marvel. This is not. That's not a flaw, that's a marvel. Now, coherence has its cost.

Coherence means that you're going to adopt one interpretation in general. Ambiguity tends to be suppressed. This is part of the mechanism that you have here that ideas activate other ideas, and the more coherent they are, the more likely they are to activate to each other. Other things that don't fit fall by the wayside. We're enforcing coherent interpretation. We see the world as much more coherent than it is.

That is something that we see in perception, as well. You show people ambiguous stimuli. They're not aware of the ambiguity. I'll give you an example. You hear the word "bank," and most people interpret "bank" as a place with vaults, and money, and so on. But in the context, if you're reading about streams and fishing, "bank" means something else. You're not conscious when you're getting one that you are not getting the other. If you are, you're not conscious ever, but it's possible that both meanings are activated, but that one gets quickly suppressed. That mechanism of creating coherent interpretations is something that happens, and subjectively what happens (I keep using the word "happens"—this is not something we do, this is something that happens to us). The same is true for perception. For Plato it was ideas sort of thrusting themselves into our eyes, and that's the way we feel. We are passive when it comes to system one. When it comes to system two, and to deliberate thoughts, we are the authors of our own actions, and so the phenomenology of it is radically different.

I wanted to give you a sense of the way I think about intuitive thinking these days, and from that point, I make a move that many people are going to be very angry with me about. Instead of talking about type one and type two processes I adopt terminology— it's not mine, but the people who proposed it have now repudiated, and have moved to type one and type two—I use system one and system two. System one does all sorts of things, and system two does other things, and I describe what is going on in our mind as a psychodrama between two fictitious characters.

System one is one character, and system two is another character, and they battle it out. They each have their own preferences, and their own ways of doing things. I will apologize ahead of time for why I'm doing this, because I don't believe that there are systems in the brain, systems in the sense of interacting parts and so on. But it turns out that our memory and our minds are shaped in such a way that certain operations are a lot easier for us than others.

A book that I read recently that I recommend is *Moonwalking with Einstein*. In that book Joshua Foer described how in one year he turned himself into the memory champion of the United States using techniques that have been around since the Greeks. The point that he makes is quite straightforward. We are very, very bad at remembering lists. We are very, very good at remembering routes. We can remember routes; we don't remember lists. If you arrange a list along your route, you're going to remember the list, and that's basically the idea.

It turns out there is something else we're awfully good at. We're good at taking an agent and assigning characteristics to that agent, and remembering that this agent has these certain habits, and it does these things. If you want to learn about system one and system two, or about type one and type two operations—really the same, think of it as system one and two—it will develop a personality. There are certain things that it likes doing, that it's able to do, and there are certain things it just cannot do, and you will get that image. It's completely crazy. There is no such thing as these two characters, but at the same time, I find it enormously useful. And it's quite funny—I'm losing friends over this. People will tell me, you're bringing psychology backward 50 years, because the idea of having little people in the mind is supposed to be a grave sin. I accept it's a grave sin. But it really helps you when you think of those characters in the mind with their own characteristics. So I'll give you a few examples.

What can system one do? I'll give you one example of the translation. What operations are conducted, automatically and without thinking? Here is one. We can detect incongruity and abnormality automatically, without thinking, and very, very quickly. And the subtly of it is amazing. To give you an example, my favorite: you have an upper-class British voice, a male British voice, saying, "I have tattoos, I have large tattoos all down my back." In 250 milliseconds the brain has a response: something is wrong.

People who speak in upper-class British voices don't have large tattoos down their back. A male voice saying, "I believe I'm pregnant," 200 milliseconds later—males don't become pregnant. This is something that happens. The world knowledge that is brought to bear is enormous—the amount of information that you would have to know and to be able to bring to bear instantly in order to know there's something incongruous here—and that happens at a speed that is extraordinary. This is just about as fast as you can pick up anything in the brain—between 200 and 400 milliseconds there is a full-blown surprise response. That's one of the things that system one does.

System one infers and invents causes and intentions. And that, again, is something that happens automatically. Infants have it. We were exposed to that. Infants can recognize intention, and have a system that enables them to divine the intentions of objects that they see on a screen, like one object chasing another. This is something that an infant will recognize. An infant will expect one object that chases another to take the most direct route toward the other—not to follow the other's path, but actually to try to catch up. This is an infant less than one year old. Clearly we're equipped, and that is something that we have inherited—we're equipped for the perception of causality.

It neglects ambiguity and suppresses doubt and, as mentioned, exaggerates coherence. I've mentioned associative coherence, and in large part that's where marvels turn into flaws. We see a world that is vastly more coherent than the world actually is. That's because of this coherence-creating mechanism that we have. We have a sense-making organ in our heads, and we tend to see things that are emotionally coherent, and that are associatively coherent, so all these are doings of system one.

Another property—I've given it the name "what you see is all there is"—and it is a mechanism that tends not to not be sensitive to information it does not have. It's very important to have a

Daniel Kahneman

mechanism like that. That came up in one of the talks yesterday. This is a mechanism that takes whatever information is available, and makes the best possible story out of the information currently available, and tells you very little about information it doesn't have. So what you can get are people jumping to conclusions. I call this a "machine for jumping to conclusions." And the jumping to conclusions is immediate, and very small samples, and furthermore from unreliable information. You can give details and say this information is probably not reliable, and unless it is rejected as a lie, people will draw full inferences from it. What you see is all there is.

Now, that will very often create a flaw. It will create overconfidence. The confidence that people have in their beliefs is not a measure of the quality of evidence, it is not a judgment of the quality of the evidence, but it is a judgment of the coherence of the story that the mind has managed to construct. Quite often you can construct very good stories out of very little evidence, when there is little evidence, no conflict, and the story is going to end up good. People tend to have great belief, great faith in stories that are based on very little evidence. It generates what Amos and I call "natural assessments," that is, there are computations that get performed automatically. For example, we get computations of the distance between us and other objects, because that's something that we intend to do, this is something that happens to us in the normal run of perception.

But we don't compute everything. There is a subset of computations that we perform, and other computations we don't.

You see this array of lines.

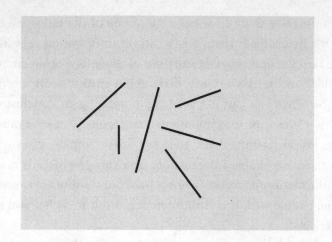

There is evidence among others, and my wife has collected some evidence, that people register the average length of these lines effortlessly, in one glance, while doing something else. The extraction of information about a prototype is immediate. But if you were asked, what is the sum, what is the total length of these lines? You can't do this. You got the average for free; you didn't get the sum for free. In order to get the sum, you'd have to get an estimate of the number, and an estimate of the average, and multiply the average by the number, and then you'll get something. But you did not get that as a natural assessment. So there is a really important distinction between natural assessment and things that are not naturally assessed. There are questions that are easy for the organism to answer, and other questions that are difficult for the organism to answer, and that makes a lot of difference.

While I'm at it, the difference between averages and sums is an important difference because there are variables that have the characteristic that I will call "sumlike." They're extensional. They're sumlike variables. Economic value is a sumlike variable.

	Set A: 40 pieces	Set B: 24 pieces
Dinner plates	8, all in good condition	8, all in good condition
Soup/salad bowls	8, all in good condition	8, all in good condition
Dessert plates	8, all in good condition	8, all in good condition
Cups	8, 2 of them broken	
Saucers	8, 7 of them broken	

Christopher Hsee, 1996

Here is a nice experiment by Christopher Hsee. He asks people to bid for sets of dishes. You have set A, and you have set B. There are two conditions in the experiment. In one condition people see this display. In the other condition, they see set A separately, or set B. Now, when you do it in this condition, the solution is easy. Set A is worth more than set B. It's got more dishes. When you see them separately, set B is valued a lot more, like $10 more than set A, because the average value of the dishes is higher in set B than in set A, and we register the prototype, and we fail to see the other. That happens all over the place.

Among other things, by the way, probability is a sumlike variable. The probability adds up over components, but we judge it as if it were an average-like variable, and a lot of the errors in probability judgment come from that type of confusion.

There are two other characteristics that I want to mention of what system one is capable of doing. It is capable of matching intensities. This is fairly mysterious how he does it, but I'll give you an example. Julie, I'd say, read fluently when she was age four. Now, I could ask you, a building of how many stories is as tall as Julie's reading capacity was at age four? And you can do it. There will be substantial agreement—that is, some people will say, "two stories," that's just not enough, and 25 stories may be too much.

We have an idea. We have an idea of what GPA matches. Interestingly enough, when you ask people to predict Julie's GPA, you say she's graduating from Yale now. They predict the GPA by matching the GPA to their impression of how well she read at age four, which is a ridiculous way of doing it, but that's what people do.

Matching intensity across dimensions is something that system one is capable of doing, and it's extremely useful because, as you will see, it allows people to answer their own question, which is what we do most of the time. It computes more than intended. The mental shotgun is that you have an intention to do something, and typically what this generates is multiple performances—not a single activity, but many things at once will happen.

Here is an example. Do these words rhyme?

Mental Shotgun

Do these words rhyme?

VOTE – NOTE

VOTE – GOAT

Here people hear the words, and they don't see them. I'm showing them to you, but the participants don't see them. "Vote note, vote goat." Both of them rhyme. People are much slower when they hear the words of a speaker that "vote, goat" rhymes than that "vote, note" rhymes. How do you explain that? Well, they have generated the spelling. They're not supposed to generate the spelling. Nobody asked them to generate the spelling, it's just in the way, they can't help it. They're asked one question, they answer other questions.

Are these sentences <u>literally</u> true?

Some jobs are snakes.

Some jobs are jails.

Some roads are snakes.

Are these sentences literally true? You spend time explaining that this is not metaphor, we don't want you to compute metaphors. "Some roads are snakes" is really slow to be rejected as literally true because it is metaphorically true. People compute the metaphor even when they're not supposed to compute the metaphor. "Some jobs are jails." People are very slow. "Some jobs are snakes" is very easy. There is no metaphor. What you get, the image that I'm trying to draw of how system one works, is that it is a mental shotgun, it computes a lot more than it's required to compute.

Sometimes one of the things it computes turns out to be so much faster than what it's trying to compute that it will generate the wrong thing.

The main things that Amos Tversky and I worked with are cases in which similarity is used instead of probability. Our most famous example is a lady named "Linda." I don't know how many of you have heard of Linda. Linda studied philosophy in college, and most people think she was at Berkeley. She participated in antinuclear marches, she was very active, she's very bright, and ten years have passed, and what is she now? Is she an accountant? No. Is she a bank teller? No. Is she a feminist? Yes. Is she a feminist bank teller? Yes. You can see what happens. She a feminist bank teller because, in terms of similarity, it's perfect to say that she's more like a feminist bank teller than she is like a bank teller. In terms of probability, it doesn't work. But what happens is that when you're asked to compute probability, probability is hard, similarity is immediate; it's a natural assessment. It will come in first, and it will preempt the correct calculation.

That was our argument. It has been contested, including even by people in this room, but that's what we proposed, and that was the general idea—a heuristic is answering the wrong question. That is, you're asked one question, and instead of answering it, you ask a related question, an associatively related question. There are other views of heuristics. Gerd Gigerenzer, who has a view of heuristics as formal operations that he calls "fast and frugal" because they use very little information, and they sometimes reach quite accurate estimates. I would contest that because the way I view this, there is no need for frugal operations. The brain is a parallel processor, it can do many things at once, and it can operate on a lot of information at once, and there may be no need for it to be frugal.

Now, what can't it do? It cannot deal with multiple possibilities at once. Dealing with multiple possibilities at once is something we do consciously and deliberately. System 1 is bound to the suppression of ambiguity, which means one interpretation. It cannot do sumlike variables. Sumlike variables demand another kind of thinking. It is not going to do probability properly, it is not going to do economic value properly, and there are other things that it will not do.

Here an important point is how you combine information about individual cases with information with statistical information. I'm going to argue that system one has a lot of trouble with statistics. System one—and here I believe the analogy from perception is very direct—it's intended, or designed, to deal with individual particular cases, not with ensembles, and it does beautifully when it deals with an individual case. For example, it can accumulate an enormous amount of information about that case. This is what I'm trying to exploit in calling it system one. It's coming alive as I'm describing it. You're accumulating information about it. But combining information of various kinds, information about the case and information about the statistics, seems to be a lot harder.

Here is an old example.

There are two cab companies in the city. In one, 85 percent of the cabs are blue, and 15 percent of the cabs are green. There was a hit-and-run accident at night, which involved a cab. There was a witness, and the witness says the cab was green, which was the minority. The court tested the witness—we can embellish that a little bit—the court tested the witness and the finding is the witness is 80 percent reliable when the witness says "blue," and when the witness says "green," it's 80 percent reliable. You can make it more precise, there are complexities, but you get the idea. You ask people, what's your judgment? You've had both of these items of information, and people say 80 percent, by and large. That is, they ignore the base rate, and they use the causal information about the case. And it's causal because there is a causal link between the accident and the witness.

Now have a look at a very small variation that changes everything. There are two companies in the city; they're equally large. Eighty-five percent of cab accidents involve blue cabs. Now this is not ignored. Not at all ignored. It's combined almost accurately with a base rate. You have the witness who says the opposite. What's the difference between those two cases? The difference is that when you read this one, you immediately reach the conclusion that the drivers of the blue cabs are insane, they're reckless drivers. That is true for every driver. It's a stereotype that you have formed instantly, but it's a stereotype about individuals, it is no longer a statement about the ensemble. It is a statement about individual blue drivers. We operate on that completely differently from the way that we operate on merely statistical information that that cab is drawn from that ensemble. So that's a difficulty that system one has, but it's also an ability that system one has. There is more along those lines, of course; it's been a long study.

But I want to talk a bit about marvels and flaws. So far, true to myself, I've spoken mostly of flaws. There is expertise, and expertise is in system one. That is, expert behavior is in system one. Most of the time we are expert in most of what we do. Sometimes it's very striking. I like the example that I pick up the phone, and my wife, Anne, says one word and I know her mood. That's very little information, but it's enough. That is expertise of a high order. How did it come about? It came about through reinforced practice, a lot of reinforcement, and a lot of practice. All of us are experts on things. The stories about fireground commanders, or about physicians who have those marvelous intuitions, they're not surprising if they have had the opportunity to learn as much about their field as I have learned about Anne on the telephone, then they're skilled. It's in system one; it comes with complete confidence.

What's interesting is that many a time people have intuitions that they're equally confident about, except they're wrong. That happens through the mechanism that I call "the mechanism of substitution." You've been asked a question, and instead you answer another question, but that answer comes by itself with complete confidence, and you're not aware that you're doing something that you're not an expert on because you have one answer. Subjectively, whether it's right or wrong, it feels exactly the same. Whether it's based on a lot of information, or a little information, this is something that you may step back and have a look at. But the subjective sense of confidence can be the same for intuition that arrives from expertise and for intuitions that arise from heuristics, that arrive from substitution, and asking a different question.

Clearly the marvels are more important than the flaws. That is in the quantitative sense, in the sense that most of what we do is just fine, and that we are well adapted to our environment, otherwise we wouldn't be here, and so on. The flaws are of some interest, and the difficulty of separating intuitions that are good from intuitions that are not good is of considerable interest, including applied interest. But that's roughly where we are.

> **A health survey was conducted in a sample of adult males in British Columbia, of all ages and occupations.**
>
> **Please give your best estimate of the following values:**
>
> **What percentage of the men surveyed have had one or more heart attacks? (18%)**
>
> **What percentage of the men surveyed both are over 55 years old and have had one or more heart attacks? (30%)**

> **A health survey was conducted in a sample of 100 adult males in British Columbia, of all ages and occupations.**
>
> **Please give your best estimate of the following values:**
>
> **How many of the 100 participants have had one or more heart attacks?**
>
> **How many of the 100 participants are over 55 years old and have had one or more heart attacks?**

I went back, because we were having that conversation almost 20 years ago, a conversation or a debate with Leda Cosmides and John Tooby, about some of our problems, and I'll mention that problem here. This is a study that Amos Tversky and I did about 30 years ago. A health survey was conducted in a sample of adult males in British Columbia of all ages and occupations. "Please give your best estimate of the following values: What percentage of the men surveyed have had one or more heart attacks?" The average is 18 percent. "What percentage of men surveyed are both over 55 years old and have had one or more heart attacks?" And the average is 30 percent. A large majority says that the second is more probable than the first.

Here is an alternative version of that which we proposed, a health survey, same story. It was conducted in a sample of 100 adult males, so you have a number. "How many of the 100 participants have had one or more heart attacks, and how many of the 100 participants are both over 55 years old and have had one or more heart attacks?" This is radically easier. From a large majority of people making mistakes, you get to a minority of people making mistakes. Percentages are terrible; the number of people out of 100 is easy.

Daniel Kahneman

There are some interesting conversations going on about ways to explain this, and Leda and John came up with an evolutionary explanation which they may want to discuss. I came up with a cognitive interpretation that is, for me, what happens. We don't have the goods on that, actually, but I'm pretty sure we're right. For me, what happens when I have 100 participants rather than percentages is I have a spatial representation. If I have a spatial representation, and all the 100 participants, or the people with one or more heart attacks are in one corner of the room, and then the people who from that set who are also over 55 years old, well, there are fewer of them in that corner of the room. For me, there is a purely cognitive interpretation of the frequency advantage, which is because speaking of individuals, of a number of individuals, calls for a different representation, as far as I'm concerned, in system one, which will enable you to work with this differently.

Yesterday at lunch, Steve Pinker and I were having a conversation that is quite interesting, about which of these modes of explanation is the more powerful. And it may turn out that it's not even a question of more powerful. Why do I like one, and why do the three of you guys like another? I thought that this might be something on which we might want to spend a few minutes because it's a big theme, the evolutionary topic has been a big theme here. And it's not that I don't believe in evolution. It must be very clear from what I've said that this is anchored. But it is not anchored in the idea that evolution has created optimal solutions to these problems, that it is not anchored in. I thought that we might want to answer any questions, or else we might just want to discuss this problem.

I'm done.

BOOKS BY JOHN BROCKMAN

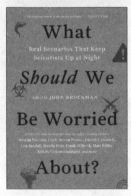

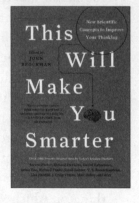

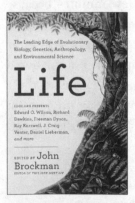

Index